A SAPPER'S WAR

HOW THE LEGENDARY AUSSIE TUNNEL RATS FOUGHT THE VIETCONG

JIMMY THOMSON WITH SANDY MACGREGOR

ALLEN&UNWIN
SYDNEY·MELBOURNE·AUCKLAND·LONDON

This edition published in 2014
First published in 2012

Copyright © Jimmy Thomson and Sandy MacGregor 2012

All rights reserved. No part of this book may be reproduced or transmitted in any form or by any means, electronic or mechanical, including photocopying, recording or by any information storage and retrieval system, without prior permission in writing from the publisher. The Australian *Copyright Act 1968* (the Act) allows a maximum of one chapter or 10 per cent of this book, whichever is the greater, to be photocopied by any educational institution for its educational purposes provided that the educational institution (or body that administers it) has given a remuneration notice to the Copyright Agency (Australia) under the Act.

Allen & Unwin
83 Alexander Street
Crows Nest NSW 2065
Australia
Phone: (61 2) 8425 0100
Email: info@allenandunwin.com
Web: www.allenandunwin.com

Cataloguing-in-Publication details are available
from the National Library of Australia
www.trove.nla.gov.au

ISBN 978 1 74331 962 8

Map of Vietnam by Ian Faulkner
Set in AIProsperall by Post Pre-press Group, Australia
Printed and bound in Australia by Griffin Press

Page 99: Lyrics from 'I Was Only Nineteen' (*A Walk in the Light Green*)
by John Schumann © Universal Music Publishing Pty Ltd
All rights reserved. International copyright secured. Reprinted with permission.

10 9 8 7 6 5 4 3

The paper in this book is FSC® certified.
FSC® promotes environmentally responsible,
socially beneficial and economically viable
management of the world's forests.

CONTENTS

FOREWORD		vii
PROLOGUE — WE MAKE AND WE BREAK ... EVERYWHERE		1
1.	THE TEAM	6
2.	THE ORIGINAL TUNNEL RATS	22
3.	TUNNELS AND BUNKERS	27
4.	SPLINTER TEAMS AND MINI-TEAMS	34
5.	ON PATROL	40
6.	THE MORE THINGS CHANGE ...	49
7.	FOOT SOLDIERS	61
8.	CORAL, BALMORAL AND BEYOND	72
9.	KNOW YOUR ENEMY	78
10.	LANDMINES DON'T TAKE SIDES	88
11.	THE SONG THAT SHOOK A NATION	99
12.	WITH TANKS ...	108
13.	BLACK SATURDAY	124
14.	CLEARING THE MINEFIELDS	130
15.	SIN CITY	139
16.	THE TETHERED GOAT	152
17.	LAND-CLEARING	165
18.	BUILDING BRIDGES	179
19.	HEARTS, MINDS ... AND A LITTLE BEAR	193
20.	THEY ALSO SERVE ...	200
21.	SAPPERNUITY	214
22.	FUNNY YOU SHOULD SAY THAT ...	223
23.	THE FINAL DAYS	239

AFTERWORD—SAPPERS SALUTED	**243**
PRAISE INDEED	**250**
THE WAR THAT NEVER ENDS	**255**
SAPPERS STILL CLEARING MINES	**261**
SAPPER SOURCES AND OTHER READING	**265**
GLOSSARY	**268**

This book would not have been possible without the assistance of many of the men of the Royal Australian Engineers regiment who were there in Vietnam and gave us their stories, their self-written recollections and permission to use accounts published in newsletters such as the Tunnel Rats' *Holdfast* and the Vietnam Sappers' *Phuoc Tuy News*. In some cases, the stories had already been rewritten or edited by those publications' editors (Jim Marett and Vincent P. Neale, respectively); in others, they were as raw as the day they were first punched into a keyboard or typewriter. Several sources also offered us material on the Sappers that they had prepared for their own books about other aspects of the Vietnam War. We thank them all.

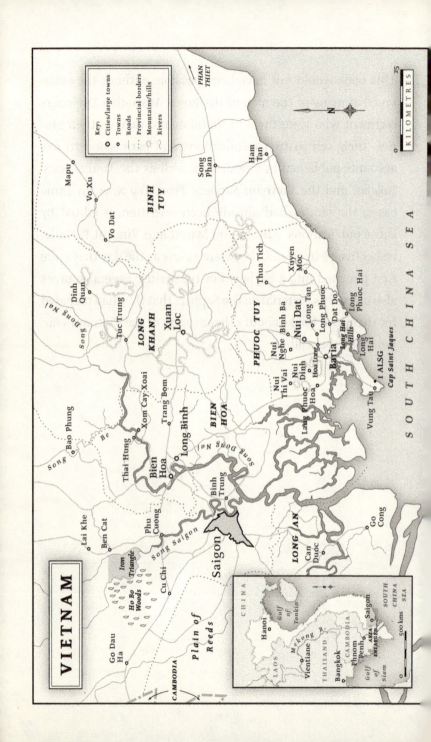

FOREWORD

You wouldn't think Sappers and Navy Divers had much in common at first glance. But you'd be surprised. The Australian Army engineers 'make and break', as their unofficial motto says. And while these guys were up to their necks in bombs, landmines and booby traps in Vietnam, their modern day counterparts, like me, are dealing with IEDs (Improvised Explosive Devices), EOD (Explosive Ordnance Disposal) and other deadly traps.

Although we obviously spend a lot of time underwater, one of my first assignments in the Navy was sailing up through the South China Sea, often close to Vietnam. I've also been on task doing what Sappers do every day. Clearance Divers have been blowing things up, rendering explosives safe, clearing bays of mines and rescuing loads of people since their inception, so I can relate somewhat to the Tunnels Rats and the Sappers who came after them. That said, there's a big difference between tunnelling and swimming underwater. In the sea you can go in pretty much any direction when things go pear-shaped. In a tunnel, you've got forwards and backwards, and that's it. On the other hand, you don't get many sharks in tunnels, except the VC kind.

I had the pleasure of launching *Tunnel Rats*, the first book by Jimmy and Sandy, and was curious to know where the story went from there. I have to say reading *A Sappers' War* took me on a journey right back to my first few hours in the forces. There's a chapter in the middle of this book about how the song 'I Was Only Nineteen' came to be written. On my very first morning in the Army, we new recruits were woken by the loud crack of machine-gun fire and then, as we lay frozen in our beds, on it came over the loudspeakers the Redgum song about a soldier in Vietnam setting off a landmine the same day man landed on the moon. It was a chilling experience; a powerful way of letting us know that our lives were changing forever and, for some of us, it might not be for the better.

So it was a pleasant surprise to discover that the Sappers' story didn't end with the Tunnel Rats—in fact, it had barely begun. I can relate to the guys who prodded and probed their way through minefields, making them safe for their mates, having spent countless hours learning the techniques. And I totally get what it must feel like to spend all day on a highly technical task only to have to saddle up as a fighting soldier when darkness falls and there are enemy to hunt down.

But there's stuff here that's totally alien to me: riding on

FOREWORD

tanks while looking out for massive landmines; building bridges at night in lashing rain, knowing the enemy is out in the bush waiting for a chance to strike; clearing jungle in a bulldozer that's a slow motion target for rocket-propelled grenades; laying 23,000 mines and then, if that wasn't stressful enough, clearing them when the high command realised they'd made a mistake.

And of course, the larrikin spirit of the original Tunnel Rats lived on, from their high jinks in the bars and brothels of Vung Tau to running a gambling den to relieve rookie infantrymen of their wages. One big difference between the blokes in the original Tunnel Rats and the men who came after them was that most of the newcomers were Nashos—National Servicemen. But for conscription, it's fair to say, many of them probably wouldn't have been in Vietnam. However, when their country's call came, they stood tall and made the most of it.

This book doesn't dwell on the politics of that situation and it's not a shot-by-shot history of the Vietnam War. But if you want to know what it felt like to be 'everywhere' dealing with a different set of problems every day, this is the closest you'll ever get to walking in a Sappers' boots.

There's one statistic that speaks volumes about the role of the Sappers in Vietnam. The engineers of 1 Field Squadron based at Nui Dat suffered heavier casualties than any other Australian unit in the Vietnam War. That says everything you need to know about this special breed of soldier.

This is their story and I salute them all.

Paul de Gelder

PROLOGUE
WE MAKE AND WE BREAK ... EVERYWHERE

> 'What is a Sapper? This versatile genius... is a man of all work of the Army and the public—astronomer, geologist, surveyor, draughtsman, artist, architect, traveller, explorer, antiquary, mechanic, diver, soldier and sailor; ready to do anything or go anywhere; in short, he is a Sapper.'
>
> —Captain T.W.J. Connolly, **the historian of the Royal Sappers and Miners, 1855**

The Australian Army Engineer was and is a unique soldier, whether he's exploring tunnels in Vietnam or defusing improvised explosive devices (IEDs) in a modern war zone. The Sappers' official motto, *Ubique*, means 'everywhere'; their unofficial motto, *Facimus et Frangimus*, means 'we make and we break'. And there, in those two phrases, we find the truth of every Sapper's fighting life.

In Vietnam, they were everywhere, literally—from deep underground in tunnels and caves to the tops of hills and mountains, from jungle tracks to rivers and the sea. As for making and breaking, they might destroy a bridge one day to hamper enemy movement, then build another elsewhere to allow for the rapid deployment of their own heavy armour. One operation could be to search for and, if required, even raze a village that had been taken over by the Vietcong. The next could be to build new homes for locals as part of their resettlement, or a school or a medical centre, as part of the Civil Affairs or 'hearts and minds' program.

However, the Australian Sappers didn't operate in isolation. They worked and fought alongside other troops, from armour to artillery and infantry, as the ingenuity and resourcefulness of a different kind of enemy forced them to change the way they had operated since the day the regiment was formed. With booby traps to be neutralised and mines to be cleared, engineers needed to be right where the action was and, critically, *when* it was happening. The men on the front line couldn't wait for engineers to be called forward when mines were encountered, booby traps were suspected or tunnels and bunkers needed to be searched. Thus, against every instinct evolved over centuries of service—during which engineers had worked in large cohesive groups—Vietnam Sappers were split into splinter teams, mini-teams and combat engineer teams (CETs) and attached to other units. Co-author Sandy MacGregor, commander of the very first Tunnel Rats, understood the profound nature of this cultural change because he was one of its instigators.

PROLOGUE

Before long, the Sapper's job was often to clear the way for other troops, whether it was ensuring the safe passage of infantry through the bush, checking caves and tunnels for booby traps and enemy soldiers or prodding the ground for mines, ahead of the armoured personnel carriers (APCs) and tanks for which they lay in wait. It wasn't unusual to find Sappers carefully leading the way *ahead* of the forward scouts.

Meanwhile, they still had all the big-picture tasks, such as constructing entire army bases, building roads and bridges, airfield runways and helicopter landing pads, and maintaining the infrastructure of the electricity and water supplies that made the bases habitable. And then there were anti-personnel mines. They had to lay them, then they had to defuse them when the enemy 'recycled' them for their own purposes. The worst, and costliest, example was the notorious Barrier Minefield—a 10-kilometre strip containing 23,000 'Jumping Jack' mines that Sappers had to lay, much against their commanders' better judgement, then had to clear when the Australian Task Force chiefs realised what a deadly mistake they had made. No unit suffered greater casualties than the Sappers in what was possibly Australia's worst military blunder since the First World War.

Our first book dealt exclusively with the men of 3 Field Troop and their status as the original Tunnel Rats. These days, the name 'Tunnel Rats' is synonymous with Australian Army Engineers who served in Vietnam—often accurately, sometimes less so. Not all those who call themselves Tunnel Rats went underground or into caves and bunkers, while non-engineers who braved the dark and deadly confines of Vietcong tunnels are denied this unofficial badge of honour. So it goes. All's fair in love and war.

However, Tunnel Rats or not, the Sappers were a special breed. They were among the first Australian forces in Vietnam, with the Australian Army Training Team Vietnam (AATTV) who trained South Vietnamese forces and eventually led them into battle. And the Sappers were among the very last Australian troops to leave Vietnam a decade later, in 1972. Some of them were bona fide Tunnel Rats, for sure; others simply carried that spirit of curiosity and courage into other areas of the war. It's no surprise, then, that the men of the Royal Australian Engineers (RAE) suffered more casualties per capita than any other regiment in the Vietnam War.

And while the story of *Tunnel Rats* focused on one small unit that made its name with one major operation, the Sappers' war is a story of infinite variety and adaptability as well as a unique quality that they call 'Sappernuity'... all combined with raw courage. In this book, we are trying to show the breadth and depth of the Sappers' involvement in the Vietnam War. So, while it is constructed with a loose chronology, beginning at the beginning and ending at the end, some of the stories are concurrent with each other and occasionally we move back and forth in time. There is a plan, however, and that's to give a sense of the Sappers' overall involvement rather than a series of dates and events.

The Australian Sappers truly had to be everywhere, and they had to make and break—but at the end of the day they

PROLOGUE

were soldiers, and thus were expected to take up arms to defend themselves and, if necessary, attack the enemy. Sappers worked with and as infantry; they teamed up with tank and APC crews, routinely saving dozens of lives while risking their own.

That's why Vietnam was very much a Sappers' war.

1. THE TEAM

As the tree-line erupted in a broiling cloud of orange and black, the boom and roar arriving a couple of seconds later, Warrant Officer John Peel watched with satisfaction. At his bidding, a brace of US Air Force Sabre jets had dropped their payload of conventional bombs and napalm on enemy positions. 'You can develop a rather big head having such awesome assets on call,' says John. 'Watching an attack requested by the ground commander and initiated by your radio call is an exhilarating experience.' This was especially true for a non-commissioned officer whose only preparation for jungle warfare had been teaching in the Sappers' small ships division in the safety of Sydney Harbour. John's previous posting had been as an instructor at the Marine Engineering Wing of the Royal Australian Engineers Transportation Centre at Chowder Bay in Sydney.

THE TEAM

How he came to be directing bombing runs and helicopter gunship strikes on the edge of the demilitarised zone between North and South Vietnam, right on the front line of the battles with the North Vietnamese Army (NVA), is part of one of the least-known stories of Australia's involvement in the Vietnam War. John, having insisted he wanted to serve in the conflict, had unknowingly volunteered for the Australian Army Training Team Vietnam (AATTV), colloquially known as 'the Team'. Its members were deployed in small two-, three- or four-man units, which were then attached to South Vietnamese forces. That's how John Peel found himself in the thick of a ferocious battle, calling in air strikes and, despite his own lack of combat experience, crossing swords with his South Vietnamese Commander.

The operation was to relieve a battalion of ARVN (South Vietnamese Army) marines which had been surrounded by an NVA force of at least two battalions, possibly more. John's unit had landed by helicopter on to a spur line leading to the mountain top on which the marines were fighting for their lives. It was a full-on battle; ammunition and grenades had to be passed forward as the wounded were carried or staggered back through their position.

'The battle continued for ages, defoliating every living tree a foot or two above our heads, and by the time I had consumed most of my cigarettes, I began to question my role in the engagement,' says John. 'I had misunderstood my mission, and this resulted in an attempt to advise the Vietnamese commander on the ground, forgetting that his people had been at war on and off for over 2000 years, that he was an experienced company commander who had been at war all his adult life. Puffed up by the recent successful direction of a set of Cobra gunships, I boldly went where angels fear to tread.

'People were dying and the wounded were returning through our position, but to this day I don't know why I crawled closer to the Vietnamese commander and said it was time he and I went forward and took control of the advance. The ARVN captain looked at me with curiosity and calmly said, "Mr Peel, I have a platoon commander up there in contact, he is in trouble but doing as well as he can at present. When he asks for help, no doubt you and I will go forward and take over. Here, have a cigarette."'

The day was eventually saved, and John and the infantry officer became close friends, but that small incident perfectly illustrates the grey areas in which the Team operated, from the first to the very last days of Australia's involvement in what the Vietnamese call 'the American War'.

When the AATTV arrived in Vietnam in 1962, its men were under strict instructions to train local soldiers but not to fight alongside them. They could plan attacks but not lead them; they could organise defences but not man them. It was not an arrangement that would survive the heat of battle—surprise attacks and ambushes far outnumbered set-piece actions—and the Team soldiers' inability to fight alongside the South Vietnamese diminished them in the eyes of their 'students'. So, typically, Team members stepped up and ignored the restriction until it was quietly dropped.

THE TEAM

Scattered throughout South Vietnam, they worked closely with and fought alongside members of the ARVN and Regional and Popular Forces—local armed militias organised for civil defence against communist fighters. They also stood shoulder to shoulder with the Montagnards, ferociously anti-communist Vietnamese mountain people who, it's said, lost half their male population in the war. Team members undertook some of the most hazardous tasks facing any Australian soldiers and became the most decorated Australian force in history, with four Victoria Crosses out of a total of 113 decorations—which also included a US Presidential Unit Citation—despite their sparse numbers compared to other regiments. Sappers would prove to be an utterly essential part of their manpower.

The longest-serving unit in the war, Team members landed long before any other Australian soldiers—including the Tunnel Rats of 3 Field Troop—and left after the rest of their comrades had gone home. Ultimately, the Team and its offshoot, the Mobile Advisory and Training Teams (MATTs), would be involved in some of the most ferocious fighting of the war. You don't win the Victoria Cross in a classroom armed with a piece of chalk and a duster.

Originally, only 30 men—15 officers and 15 warrant officers and sergeants— joined Colonel Ted Serong, in Saigon in August 1962. From such modest origins, their numbers grew to a peak of over 200 in 1970 and, by the end of the war, almost 1000 Australians and 11 New Zealanders had served with the AATTV or MATTs. Sappers were soon right there with them, as they would be for the duration of the conflict.

In the early days, the AATTV evolved into a special-operations unit by any other name, deployed right on the front lines, often unofficially and sometimes unaccountably. Usually working alone or with American advisors, some AATTV members were involved with the US Central Intelligence Agency's covert operations. Many Team members were even involved in the Phoenix Program, which targeted senior Vietcong members for assassination. These guys were much more than trainers—they were inspirational leaders showing the way with their military know-how.

The first Sapper to become part of the Team, Warrant Officer John Swanson, arrived in June 1963; he was joined in October by Captain Rex Clark and Warrant Officer Reg Collinson. The fact that there was no one below the rank of sergeant in the original Team shows that they were, indeed, an elite force.

Sappers became increasingly heavily involved in this 'ghost squad'. Its members mostly operated as individuals or in pairs; often, like John Peel, they directed artillery and aerial fire support from close quarters. Even before Sandy MacGregor's 3 Field Troop left Australia, reports filtered through from Vietnam about tunnels—albeit short rat-runs from houses rather than the huge complexes they would later discover—and booby traps. These would certainly have come from Team members.

A Team member was among the first Sapper casualties in Vietnam. Sergeant James MacDonald died from gunshot wounds on 7 February 1966, while attached to the AATTV,

THE TEAM

barely a month after Corporal Bob Bowtell became the first Tunnel Rat to die. By the time the Australian Army left Vietnam, 70 Sappers, none below the rank of corporal, had passed through AATTV ranks. Most had been selected for the Team because of their rank and their specific skills, either as engineers or as trainers or both. This took many well beyond their comfort zones, into roles they often felt ill-equipped to perform. Out of necessity, however, most would learn quickly on the job.

John Peel recalls an overwhelming sense of confusion and bewilderment as he went from the relative safety of Sydney Harbour to the heart of a war zone: 'The major who interviewed me at the end of the Tropical Warfare Advisors course at Canungra said, "Mr Peel, you have never even served in a field unit—what use could you be to us?" I responded, "Sir, infantry tactics are adequately explained in the training pamphlets and I can read." I showed him the statement "Understands Battalion Tactics" from my course report but he noted my "demonstrated inability to throw hand grenades accurately". Finally, he said, "You really do want to go, don't you?" My answer resulted in me being on a bus heading for the Free World Headquarters in Saigon.'

John remembers the culture shock when a large US Army bus rolled onto the tarmac at Tan Son Nhat airport with a warrant officer leaning out the front door. 'About twelve or more of us disembarked the Pan Am 707 and, sweating profusely after a short walk, boarded the bus for the trip to AATTV HQ in Saigon. It was as if someone had opened an oven door in which rotting fish soaked in aircraft fuel was being baked. I flopped down next to Harry Buckley, a fellow Sapper. Our escorting warrant officer, unaffected by the heat and neatly dressed in jungle greens and slouch hat, stood to welcome us. The second name to be read was mine. "WO2 Peel, 2nd Battalion, 1st ARVN Regiment at Quang Tri." I was transfixed and never heard the rest of the posting list.'

John Peel couldn't have known it but he was in the presence of a Sapper legend. Harry Buckley, aka Black Harry, would become an iconic, almost mythical figure among Aussie engineers. In any case, John was too busy coming to grips with his new assignment. His posting was to 1 Corps, which had been the centre of combat activity since 1962. This was where the Australian Team members had created something of a legend of their own, fighting alongside and leading South Vietnamese forces.

Even more alarming for the newly arrived Warrant Officer Peel was his commanding officer's warning that a stated aim of the enemy was to capture an Australian, preferably an advisor, for bartering purposes at the Paris Peace talks. 'My mind was dominated by the thought of avoiding capture, and/or what I would do if capture was inevitable,' John recalls. The men were handed briefing notes on the areas where they would be serving. 'I read about Ray Oliver with the 4th Battalion being awarded a Military Medal, and how Fire Support Base O'Reilly (in his new Tactical Area) had experienced almost unbroken contact for the previous three months. The 2nd Battalion report described the aggressive defence of O'Reilly and the American 101st Airborne Division fire base Ripcord.'

THE TEAM

It was a truly daunting assignment. John and the other new arrivals were kitted out at an American quartermaster store before flying out the next morning. On arrival in Da Nang, John discovered that four loaded magazines in the pockets of his backpack had been 'stolen'. Welcome to Vietnam, he thought, but he later learned that live rounds weren't allowed on domestic flights; they had probably just been confiscated. The Team's HQ for I Corps was a French-style villa owned by a Vietnamese widow and her daughter, who lived on the premises and provided light meals and housekeeping.

John had been told he was replacing infantry Warrant Officer Dave Powell, and he desperately wanted to talk to Dave before he caught his plane home. He discovered him leaning on the bar at 'Uc house', where a sign proclaimed *Nuoc Mam Hall—Home of the Expendables*. 'Dave was in animated conversation with three or four others, and the message was clear—we would have time in the morning for a briefing,' says John. 'At 6am the following morning, the administration warrant officer shook me awake and said I had less than an hour to get out to the airport for a plane ride to Quang Tri. I never did see Dave Powell again and so I knew nothing about the situation I was being rushed into. Later I found this had been a common experience.'

The Military Assistance Command Vietnam Team 3 (MACV3) team comprised Australian Infantry Captain Ernie Martens, who was senior advisor, Captain Jack Jandouski of the US Marine Corps, US Infantry Staff Sergeant Jimmy Miss, and John. The team usually deployed in pairs—one officer and one senior NCO—on weekly rotation. Each man looked after one command post, which meant they usually worked alone.

The 2nd Battalion of the 1st Regiment (ARVN) was commanded by Major Lee-Kach-Kha, whom John describes as a competent and experienced commander who held Australians in high regard. In all, 16 Australians had served with the battalion, which was still recovering from the death of Warrant Officer Jack Fitzgerald, who was killed in action earlier in 1970 during a combat assault with the regiment's 1st Battalion. In combat, the Team advisors covered any shortfall in ARVN command and control, provided artillery and tactical air support, and liaised with helicopter transport, gunship control and medical evacuation units.

John Peel was the last Australian to serve with the 2nd Battalion of the Republic of Vietnam Army Regiment and, despite his less than auspicious origins as an engineer-sailor, his contribution was acknowledged with the Republic of Vietnam Gallantry Cross with Silver Star, plus a US Army Commendation medal for Meritorious Service. He retired from the army as a major, a long way from that raw warrant officer sitting stunned in Saigon, learning about his 'fishing line to firing line' career path.

The demand in the AATTV and Mobile Advisory and Training Teams (MATTs) for Army Engineer NCOs led to many strange transformations. From teams exclusively comprising officers and sergeants in the original AATTV, the MATTs evolved into small units usually consisting of two warrant officers and three or four corporals, one of whom was usually

THE TEAM

a Sapper. The need for engineering NCOs meant men were drafted from the furthest reaches of the Sapper ranks—all they required, it seemed, was a stripe or two on their arms to show they had experience and leadership skills. 'The corporals would always have one infantryman and an engineer,' says Peter Aylett, who completed his first Vietnam tour as a plant operator, or 'plantie'; when he came back for his second stint, from September 1970 to 1971, he found himself attached to the MATTs. 'The other two would probably be an artilleryman, a medic or someone from another corps,' Peter says. 'It was a pretty sound arrangement.'

Going from driving a bulldozer to leading local militias in combat actions against the Vietcong was such a huge change in his duties that Peter didn't see it coming. 'It's got me beat how I ever finished up in it,' he recalls. 'I had applied to go back to Vietnam but I never knew much about AATTV, and it wasn't until I started doing all this infantry-type stuff and learning about every bloody weapon under the sun that I realised I wasn't going over there as a bulldozer driver or any such thing—I was going to be a field engineer. I thought, "You'd better brace up a bit, Pete, and learn a bit about your basic stuff—mines and booby traps and the like—otherwise you're gonna get your head blown off here.' It wasn't a game for novices. I suspect that I was probably put on the training team because someone wanted to get rid of me,' he laughs. 'That was the best reason I could come up with.'

Peter was posted to MATT 9 at Phuoc Loi. 'If you drew a line down the road between Dat Do and the tip of the Long Hai mountains, or Lang Phuoc Hai, it was about halfway. We were with a Vietnamese company there on the southern side of the township, to train the company, to go out on patrol with them and ambush with them. So it was more so than an engineer's role—it was pretty well a basic infantry role.' Peter was there for six months but moved on to MATT 3 in Long Dien after what he describes as a 'personality clash' with a new warrant officer.

'MATT 3 was the only one that had enough Regional Forces companies to form a battalion, so we were a bit bigger than usual,' he says. 'We had a captain or a major in charge, which changed over a few times when we were there, plus three warrant officers and three corporals. I felt quite comfortable always carrying the radio. I would be the one to call in the gunships. On a couple of occasions we had battleships going past and I called those in—it was quite an experience. We also had an observation post up in the Long Hais, and I was up there for a total of about two months on three- to five-day rotations with the other members of the team.'

Peter recalls that training and advising was still part of the job description. He ran a mine course at Phuoc Loi and taught it for all the companies supported by MATTs, but he felt there was a potential problem with 'over-advising'. 'You give people too many rules and so they remembered none,' he says. 'So you were much better off just picking the main point that you would like to instil and always enforce it and lead by example—that sort of thing.

'I had two major rules. The first one I stomped on pretty

quick—if you find a bloody mine, come and get me and take me to the mine; don't bring the bloody mine to my hutchie. The word got around pretty quick—I don't think I left any question of doubt about what I meant.'

A hutchie, also known as a hootchie or bivvy, is the Diggers' bivouac or sleeping shelter which can be as elaborate as a permanent tent while in camp, as versatile as a hammock between trees while based at a fire support base (FSB), for instance, or as basic as a scrape in the ground while out on patrol.

'My other main rule was "spread out",' Trevor recalls. 'I used to say, in my best Vietnamese, "One M16 mine equals one Aussie dead, two Aussies wounded. But for Vietnamese, one M16 equals 20 Vietnamese dead,"' he laughs. 'But would they listen? You'd see us on patrol, it would be bunch, bunch, bunch, then a six-metre gap, then me and then everybody else right up against my backside.'

Disregarding the rules could have fatal consequences. 'One bloke that I went out with quite regularly, Bob Hunnisett, was always saying, "Never use the same path twice and don't set up routines." But they ignored him and took the same route up to the top all the time.' The cost, when it came, was, as Peter describes it, heartbreaking: a very well-organised ambush saw almost half the patrol killed. 'We were doing a rotation, going up with another platoon, 16 of us—including me and Bob—and they sprung this ambush,' recalls Peter.

'It was very, very well done—you've got to pay the Vietcong credit for that. Everywhere these blokes dived, they had an explosive device, command-detonated. As soon as the blokes got behind something for cover... boom! They went to cover right on top of these things and we lost seven out of the 16 of us. Our blokes were pretty well pinned down by a spasmodic machine gun until they finished doing what they wanted with the bodies.'

Peter's recollection confirms the sense that the Team and the MATTs were actually directing these local units rather than simply training them. 'The only real training, as you would imagine it to be, was when they created a training school in Nui Dat. Beyond that, all training was done in the company lines in the field. You went to the field with them and led by example. There was no sitting down, listening to your instructors and then going out fighting in a real war. You had to be out there with them.'

Peter is convinced that was happening in the AATTV, right from the beginning, despite the ban on engaging the enemy in combat. 'If they weren't leading initially, which I don't believe is the case, it's because they had someone sitting in the seat and he was taking directions from his adviser, telling him what to do and how to do it. If that's not leading, I don't really know what is.'

Alex Skowronski had a similar background to Peter; originally with 17 Construction, he too returned as a MATT member on his second tour: 'The make-up of the team was two warrant officers, one a senior one and team leader,' Alex recalls. 'Then you had an infantry corporal, who was basically responsible for training the local forces, a weapons specialist for heavy

THE TEAM

machine guns and so on, a Sapper for the engineering aspect, and a medic.

'South Vietnam was broken down to 1 Corps, 2 Corps, 3 Corps and 4 Corps—1 corps being the northern part, 2 Corps being predominantly the highlands, 3 Corps the centre, and 4 Corps the Delta. You had warrant officers who were basically advisers at battalion levels. I may be corrected here, but they were also responsible for making sure the South Vietnamese soldiers got their pay and all sorts of stuff. I like the Vietnamese but there was a fair amount of corruption involved. Some battalion commanders would pay their soldiers in rice rather than money.'

And it wasn't just corrupt Vietnamese officers they had to be wary of. One of the duties of the MATTs and their Vietnamese platoons was to protect local villages from incursions by the Vietcong. The Popular Front troops that the Aussies worked with were basically just local armed militias based in the villages. 'Within some of the units you had Vietcong, there was no question about that. Ray Deed and I, Bill Parry, Snowy Ray and Mick Dawkins were working for quite a period of time with the Popular Front Platoon at Nai Xao and we knew there were a few in there—we just didn't know how many.

'We went into the village one afternoon—there was Ray, Bill, Mick the medic, me and Mik the Vietnamese interpreter. Mick and I were talking to the platoon commander, while Mik was standing about 15 to 20 metres away. I looked around and one of the Vietnamese soldiers was giving Mik some stick. Nothing happened that night but there was tension. The next morning, after we left there, Ray Deed wasn't happy about the setup and said, "We are going to go back tonight." He told me to go and tell Mik to get some sleep 'cause we were leaving. I went downstairs—we lived in an old French villa. I told Mik and he said, "I'm not going." I said, "What d'you mean you're not going? Was it that discussion you were having with that soldier?" He said, "They are going to kill me if I go back." So I went and told Ray, who said, "Okay, well get someone else. Give Mik a break."

'We got this Vietnamese signalman we used to call Gizza— 'cos he was always saying "gizza piece" of this or that—and we took an extra Vietnamese sergeant. We headed back late that afternoon, got into the village and prepared for the night activities. It was well after dark when we checked the defences and all that sort of stuff. There was a track going out of the back of the village. Ray asked the chief and platoon commander what they had out there on this track, meaning what night defences.

'They told us they had grenades on trip-wires. I whispered to Ray, "That's bullshit," and I said I'd check it. There were trees everywhere. I got a long branch and stripped the leaves— I was flicking it from the ground to up above my head, feeling for the trip-wires. I went about 40 metres but there was nothing there. It meant their Vietcong mates could come walking in and out. Things got a lot more tense after that, and Ray said, "The first opportunity we get, we are out of here."' But that wasn't the end of the story. 'About a week after that, the Vietnamese National Police Field Force came and took nine of the 20 guys in this platoon away.'

THE TEAM

This was not an isolated incident. Not being entirely sure who was friend and who was foe—especially in the Regional and Local Defence Forces—was a problem that the Australians constantly faced, especially after they were given their own area of operations in Phuoc Tuy.

2. THE ORIGINAL TUNNEL RATS

The period leading up to Australian forces being given their own tactical area saw the biggest combined American and Australian operation of the Vietnam War. Operation Crimp could and probably should have changed the course of history. The discovery and investigation of the massive Vietcong tunnel systems in the Iron Triangle, north of Saigon—comprehensively covered in our first book, *Tunnel Rats*—was cut short before the large numbers of North Vietnamese and Vietcong forces hiding there could be flushed out. However, more tunnels would be discovered and searched throughout the war. But as soon as they were investigated and blown up, more would be dug to replace them. In fact, tunnels at Cu Chi in the Iron Triangle would ultimately be used as a base for the final assault on Saigon and the fall of South Vietnam.

THE ORIGINAL TUNNEL RATS

Back in 1964, the Australian government had decided to increase its support of its American allies. Rather than merely providing 'advisors'—although we now know they were a lot more than that—it resolved to send substantial forces with clear military missions. So, while plans were in place to conscript and train an army of national servicemen (or 'Nashos'), in June 1965 the 1st Battalion, Royal Australian Regiment (1 RAR) was dispatched to serve alongside the US 173rd Airborne Brigade in Bien Hoa province. A specialist unit of engineers, 3 Field Troop—under the co-author of this book, Sandy MacGregor, then a young captain—was sent to help deal with unconventional challenges such as booby traps and tunnels, as well as fundamental engineering tasks such as building and breaching bridges and preparing the way for the arrival of the Australian Task Force (ATF)—including the first of the Nashos—the following year.

But before they took on that massive logistical task, in January 1966 there was Operation Crimp, during which 3 Field Troop made its mark as the first Allied soldiers in the Vietnam War to go down and properly investigate the Vietcong's major tunnel systems, creating the legend of the Tunnel Rats. Crimp was undertaken because the Allies' military command rightly thought the area of the Iron Triangle might be the location of the Vietcong's southern command headquarters. However much they bombed the area, renewed high levels of enemy activity would soon be spotted. A plan was hatched to drive the Vietcong out once and for all by entering the area from the south and pushing north, where Australian forces would be waiting for the retreating enemy—hence the codename Crimp.

The operation involved some 8000 troops from the US and Australian ranks. It was the largest American operation in Vietnam to date, and one of the biggest in the war involving Australian troops. Apart from the fact that the plan was leaked and a trap set by the Vietcong—which was averted only when Australian commanders changed the landing site and the launch date at the last minute—there was another surprise for the Allied troops. The enemy weren't running off into the bush and the waiting guns of the other forces. Instead, they would drop from sight and reappear elsewhere, often at close quarters. It was deadly close-range sniper fire from what seemed to be an anthill that first alerted Sandy's men to the fact that there were underground bunkers. Further investigation revealed a virtual underground city, with tonnes of weapons captured and reams of documents discovered.

The wealth of equipment and intelligence discovered by 3 Field Troop exposed the inadequacy of the standard practice employed by American forces up to then: smoke would be blown down the tunnels to flush out any enemy troops and reveal other entrances, which would then be destroyed. After massive amounts of munitions and documents were pulled out of the tunnels by Australians, general orders to all forces were changed to ensure that tunnels were properly investigated before being destroyed. According to the seminal book on this operation, *The Tunnels of Cu Chi*, the Americans only went down the tunnels the day after Sandy MacGregor's men became the very first Tunnel Rats and demonstrated the value of properly investigating the tunnels.

At this time, apart from the AATTV, who worked more with South Vietnamese forces than their own countrymen, Australian soldiers operated within a relatively contained area based around Bien Hoa, just a few kilometres east of Saigon, with operations taking them as far as the Mekong Delta, to the west. In 1966, Australia's chief of army, Lieutenant General John Wilton, proposed an Australian task force that would operate independently of United States forces, allowing Australian soldiers to fight the war the Australian way.

This wasn't just nationalistic hubris. At the risk of overgeneralisation, the American battle philosophy was to be loud and aggressive and invite the Vietcong to engage, believing the US Forces' vastly superior firepower would win the day. In stark contrast, many of the regular army Australians were trained and experienced jungle fighters. Their skills had been honed in Malaya and elsewhere in the region and would be passed on to the Nashos when they arrived. Their approach was subtler; in many ways, it was more like that of the North Vietnamese and Vietcong, who had sharpened their tactics over decades of guerrilla fighting against the French.

It's worth noting that some Americans didn't see the Australian methods as in any way superior; in fact, to some the idea of sneaking around trying to outwit the enemy was seen as cowardly and underhand. Many Australians admired the American troops' bravado and gung-ho attitude but they also took great pride in their bush skills and would not have had it any other way.

The expansion of Australian involvement in Vietnam was

A SAPPERS' WAR

approved in March 1966, and Phuoc Tuy province—to the south and east of Saigon, on Vietnam's southern coast—was selected as the Australians' zone of operations. There were hills and mountains but much of the province was flat and used to grow rice and rubber. The port of Vung Tau provided naval access and could serve as a logistics base. The South Vietnamese government's control over the province didn't extend far beyond the provincial capital, Baria, and the Vietcong had an extensive and powerful organisation throughout the towns and villages. Their strong and ultimately unassailable presence in the Long Hai mountains would present a constant and frequently fatal thorn in the ATF's side.

Thus, to a great extent, the ATF was operating in enemy territory. Ambushes were commonplace on the roads of the area, and some could only be used with well-armed escorts. Landmines and anti-tank bombs were used to devastating effect. The number of communist forces operating in Phuoc Tuy has been estimated at 5000, and they were supported by many of the province's villages. Essentially, for the rest of their time in Vietnam, Australian forces would be locked in a battle with the Vietcong for control of Phuoc Tuy.

Needless to say, Sappers were hugely influential in the campaign. From building the ATF's bases at Vung Tau and Nui Dat to winkling the Vietcong out of their underground havens, the Australian Sappers experienced every aspect of the war and knew their enemy almost as well as he knew himself.

3. TUNNELS AND BUNKERS

Vietnamese today refer to the conflict as the 'American War'. But for the residents of Phuoc Tuy province, centred on Binh Ba with the harbour city of Vung Tau in the south and Bien Hoa in the west, it could just as easily have been the *Ucdaloi* or 'Australian War'.

After Operation Crimp, 3 Field Troop went on to prepare an area of the military base at Vung Tau for the arrival of the Task Force, including the first of the Nashos. They then shifted their attentions to building the massive base at Nui Dat, which would become home for the Australian forces for the rest of the war. Three Field Troop was eventually absorbed into 1 Field Squadron, becoming simply 3 Troop, but the Tunnel Rats' story had only just begun. There were many more tunnels to discover, and now they would all be explored. And even though those who followed had a better idea of what to expect, there were still plenty of surprises—if not shocks—awaiting the next men to go underground.

Lieutenant George Hulse, who was there from 1968 to 1969, vividly recalls the challenge of entering and searching tunnels. 'We would check the entrance for mines or booby traps by prodding the ground,' he says. 'Then, using torches, we would shine intense white light on the first two metres of tunnel. We looked for three signs. If there were a thousand tiny silver eyes looking back at you, they were tiny spiders and their webs were intact. That meant the hole was probably "cold"—that is, no one was in there—so a few risks could be taken with the speed of the search.

'If the spiders were there but many of the webs were broken, the hole had probably been occupied recently, and this meant a slower search was required. If there were no eyes and no webs, and there were scratch marks on the wall where someone had passed, then it was considered a "hot hole" and searching was done with extreme care and vigilance. Also, if there was the distinctive odour of other humans down there, it was a candidate for aerosol explosive charges to be pumped in, although that was sometimes used even without the tell-tale signs. Of course, if somebody actually saw an Asian person disappear down the hole, it was automatically a hot hole.'

According to George, Australian Sappers soon discovered that they needed an additional limb to be able to operate effectively. 'When operating down the tunnels, sappers needed three hands. One for the 9-mm Browning pistol, a second for the prodder—in the early days we used bayonets, although we probably shouldn't have—and a third for the torch,' says George. 'I realised the need for a special tunnel torch after I had a contact with a Vietcong "caretaker" down a tunnel in late 1968. The torch of favour was the nine-volt Dolphin with its brilliant white light. In any contact, you had to very quickly change hands between pistol, prodder and torch.'

It's worth noting that the men of 3 Field Troop had requested miners' head lamps for this very reason; a batch had quickly been sent out by BHP, the mining and steel-making company. But these had some considerable disadvantages, not least that they provided a very convenient aiming sight for enemy troops, and they didn't illuminate the target if you happened to be looking in a slightly different direction from where you were shooting. Rifles, of course, were impractical because of the restricted space but even pistols had their drawbacks.

'When you fire a 9-mm pistol in a very confined space,' George says, 'there is a boom that drives your eardrums into a high-pitched scream, a flash that blinds you temporarily and then an incredible amount of choking smoke from the discharged round. You create your own sudden "battlefield obscuration" in a tiny space. And to make sure that the enemy soldier stays hit, you have to blindly fire several more rounds down the tunnel.'

The need for a pistol-mounted light became evident when George was in a tunnel with 'Fingers' Murray of 1 Troop and they encountered an enemy soldier armed with a captured

US-issue 0.45-calibre Browning automatic handgun. George let rip with his pistol before the enemy was able to fire. 'When I fired my eighth round, we assured ourselves that the enemy soldier was dead,' he says. 'Then, in the smoke, we switched off our torches and snuggled up to each other for mutual reassurance, as we were both shaking from the adrenaline. It was impossible to talk due to the ringing in our ears, but that did not seem to matter at all. We each experienced a chill and drew warmth from the other's company.

'When that sensation passed, the shaking subsided and we pulled the dead enemy soldier out of the tunnel. It was a very hard job. It was hard to get hold of him and pull him in a straight line without him jack-knifing and jamming against the tunnel wall. When we finally got him to the surface, we realised that he was only about thirteen years of age. Not only that, his pistol was totally rusted—he was never going to fire that weapon. It had the safety system on and it was not "cocked"—there was no bullet in the chamber. Fingers and I were devastated; had we known, we could have captured him alive. The infantry platoon commander, who was present when we brought the body out, was equally distraught at the age of the boy and walked away in disgust. The boy may have been enemy, he may have had an intent to kill us, but we were in no mood to celebrate. It was a very unhappy experience.'

On his return to Nui Dat, George started a competition among 1 Troop to come up with a design for a light which would point in the same direction as the pistol, and which would not fail to show good light down the tunnel at the same

TUNNELS AND BUNKERS

time as the weapon was fired. It would also need to allow the bearer to change magazines very quickly without the torch getting in the way, in case there was more than one enemy soldier. Finally, the torch would have to be removable from the pistol when anything was discovered in the tunnel that merited a closer look.

'We were unable to come up with a light that could be fitted to the pistol and continue to illuminate the area of vision after firing the pistol,' George says. 'Even the inventive "Gyrogearloose" Sapper Bruggemann, with his fertile imagination, could not come up with a workable device. Many designs were tried in 1 Troop but all failed.'

However, Sapper Jim Duffield, the squadron's storeman, applied some lateral thinking and few days later he had the answer. Jim had acquired a battery-powered miner's head lantern, with high and low beam, connected to a belt-mounted battery pack. There was nothing exceptional about this—miners had been using similar arrangements for years. But miners didn't need to carry guns. 'The genius of this invention was that Jim attached the headlight to a large spring-loaded handle, which exactly fitted the pistol grip of the handgun,' recalls George. 'The light could be snapped on and off the pistol grip at will, to help in illuminating the work at hand, and could then be snapped back onto the pistol for further movement in the tunnel. You pointed the pistol and got white light exactly where the muzzle was pointed. And you could fire the pistol without the light going out, or the beam straying from where you wanted lots of light. You could also change magazines without the need to disturb the light. I was delighted with Jim's tunnel torch.'

George encouraged Jim to put his invention up to the Army Suggestions Committee for eventual mass production, and for the royalties and recognition that should have rightly gone his way. But the Army Suggestions Committee told Jim that they already had the idea under development, thanked him for his contribution to the army's combat efficiency, and then gave him a cheque for $8 as a reward for his efforts. 'I served for 32 years in the army, and to this time I am still unaware of anything faintly resembling Jim's tunnel torch,' says George. 'It appears that the idea went into the "graveyard of good ideas" under the Army Suggestions Scheme that existed at that time.' Another good idea that found its way to the too-hard basket was fitting the pistols with silencers. It was decided that the silencers could too easily find their way into the hands of criminal types, and so that idea was kyboshed too.

There were other underground challenges awaiting Sappers: bunkers. These were deep defensive pits and bunker systems varied in size and complexity. Sometimes they were in twos and threes, but often they were dug in larger complexes of 20 or 30 or more bunkers, linked by tunnels and trenches. When they were discovered, the skills of the Sappers would once again be to the fore, explains Jim Marett, a Sapper with 2 Troop in Vietnam from June 1969 to June 1970 and now editor of *Holdfast*, the Australian Tunnel Rats' online newsletter.

'The average bunker was 2 metres wide by 3 metres long and about a metre and a half deep,' writes Jim. 'Their overhead protection usually consisted of logs with earth packed around them. The outside of the roof was covered in packed earth and camouflaged with reed matting or leaves. There were sometimes two entrances into each bunker, enabling rapid entry or exit if required. The soil in the region was perfect for the purpose. Straight walls with sharp corners could be created, without crumbling or falling away of the soil. After a relatively short time, the walls of the bunkers "seasoned" and in many cases looked remarkably like concrete rather than the red laterite clay in which they were created.'

The Sapper's job was to search the bunkers and then set them up for demolition. Before entering each bunker, he'd carefully check for mines or booby traps, usually in the track close to a bunker's entrance, or grenades connected to tripwires in the opening leading down into the bunker. Then he'd look for enemy weapons, documents and stores, and at the same time see if there were trapdoors leading to tunnels or adjoining rooms.

'There would be varying amounts of stuff left behind, depending on how rapidly the VC had vacated the base camp,' recalls Jim. 'It was a unique experience to sift through the enemy's stuff. It really brought it home that they were real and they were out living in the bush, just like we were.'

Between tunnels, bunkers and booby traps, warfare for the Sapper in Vietnam was rapidly evolving and the whole philosophy of how engineers went about their business was changing.

4. SPLINTER TEAMS AND MINI-TEAMS

In a clearing in some low scrub next to the parade-straight rows of trees in a rubber plantation, infantrymen stand as still as their nerves will allow. To be motionless is almost impossible but to move could be fatal... to yourself and your mates. In 90 per cent humidity and 40 degree heat, sweat pools in the small of your back like a blocked storm drain. You try to stay cool but the knowledge that you have walked unwittingly into an enemy minefield can only add to the perspiration. Only two men move: the Sappers. The first picks and prods his way forward, using his bayonet the way a dentist wields a probe when you've got his nuts in your hand... very, very carefully. The Sapper's offsider is a dozen paces behind him; the grunts' forward scout is several metres further back again, scanning the tree-line for any signs of a sniper.

SPLINTER TEAMS AND MINI-TEAMS

The number one Sapper finds the first mine, and marks it. Now his feet are also glued to the ground. Only his head moves as he scans the undergrowth for another tell-tale set of landmine prongs. Or it could be a trip-wire leading from a seemingly innocuous Coke can, in which the ubiquitous soft drink has been replaced by explosives and packed tight with nails and ball bearings. The enemy knows that one anti-personnel mine can be found; and they know that moment of fright followed by elation can lead to deadly carelessness. If a mine is triggered, a second device will get anyone rushing to the aid of an injured comrade. A second chance to die is always left for the unwary. The booby trap is spotted—a fishing line trip-wire attached to a hand grenade in a tree overhead. Now the Sappers can get to work, defusing the mine and grenade, finding the other mines and marking a path so the infantry patrol can advance.

Elsewhere in Phuoc Tuy province, a Sapper smacks down hard on an APC commander's helmet. Sitting on top of the vehicle, the engineer has spotted clues that few others would have seen: a patch of earth in the road a marginally different colour from the area around it; an X formed from long blades of grass on the verge; a couple of centimetres of electric wire exposed in a rain puddle. The bash on his helmet sees the APC driver slam on the brakes. The Sapper gets down from the vehicle to investigate and discovers a bomb that's big enough to blow him, the APC, its crew and its human cargo to kingdom come. He can only hope that the bomb is attached to a pressure switch rather than set to be fired by a remote trigger held by an unseen enemy somewhere in the bushes. His mate unpacks and assembles the mine-detector they've strapped to the side of the APC. His job will be to sweep the road ahead for more unpleasant surprises so the convoy can carry on ... once the big bomb has been neutralised, of course.

Those two scenarios were the day-to-day work of an Australian Sapper in Vietnam. And, as if that wasn't challenge enough in itself, there was no handbook to work from, no tried and tested set of instructions. The game had changed profoundly for Sappers and they were, to a large extent, making up the rules as they went along. Before the Vietnam War, engineers always tried to work together as a group. They were more effective that way, getting the job done for the best result in the shortest possible time.

In some contexts, that makes perfect sense. The 'job' could be huge infrastructure tasks, from building roads and bridges to destroying them; it could be digging wells or gouging out quarries to provide materials for other tasks. It could be constructing an airfield, a helicopter landing pad, an army camp or a village. It could be setting up an electric generator or a drainage system. And, just as often, it could be taking all of those things apart—carefully bit by bit, or violently with an explosives charge.

All of these tasks and more confronted the Australian Army Engineers in Vietnam, and they needed to be done well and on time. You don't have time to lean on a shovel when a convoy is just a day away and you have to get it across a river or risk having it stationary, a sitting target for the enemy, stretching literally for miles through the jungle. And all the time there's a chance you could be shot at and have to return fire.

SPLINTER TEAMS AND MINI-TEAMS

The war in Vietnam didn't change that, but it added a new range of Sapper tasks to their repertoire of skills: finding and 'delousing' booby traps, and, of course, searching and clearing tunnels and bunkers. Suddenly, it made no sense to always deploy Sappers in large units. These highly localised dangers were immediate and potentially deadly—infantry, armour and artillery couldn't wait for a Sapper to be kitted up and sent out to them.

Wherever they went, infantry and armour encountered threats best dealt with by engineers, and it soon became clear that the most effective use of Sappers was in small teams of two, attached to other units. These would be called 'splinter teams' when they were with infantry and 'mini-teams' (basically the same, only carrying a mine-detector) when they were with tanks and APCs. Combat Engineering Teams of six or more men were held on stand-by for bigger operations such as blowing up a large tunnel or bunker system. Sandy MacGregor was among the first to embrace this new way of deploying Sappers.

'[This] was a period in which infantry/sapper cooperation was first developed in Vietnam,' writes Brigadier P.J. Greville in his book *Ubique: The Royal Australian Engineers—Volume Four*. 'In each operation in which 3 Troop participated, it was faced with new problems to which MacGregor and his men applied their basic skills. In doing so they evolved practices which became the basis for further development of Corps doctrine in Vietnam. Maj W.W. Lennon, the first OC 1 Field Squadron, wisely used the remaining time of 3 Troop in Vietnam, to ensure that the local knowledge of this troop was transferred to his squadron for use in Phuoc Tuy.'

It was this necessary evolution of the Sapper that laid the foundations for how they work in war zones, right up to the present day. 'In my time I used a combat team per company,' says Sandy. 'It consisted of one corporal on Company HQ and three splinter teams of two Sappers, each team being with a platoon. At battalion headquarters I had combat teams ready to support splinter teams if a job was too big for them. These combat teams were variable in size. As time went on, in Phuoc Tuy province the mines became a real issue, so that a mini-team was placed with armour—both tanks and APCs. The major difference between a splinter team and a mini-team was that the latter had a mine-detector. You see, the mini-teams travelled in luxury—ready to be blown up on an APC at any time,' he jokes.

Ian Kuring, in his paper 'The Australian Army and the Vietnam War, 1962–1972', explains the change in thinking: 'The prolific use of mines by the Viet Cong in the southern areas of Phuoc Tuy Province and the high number of battle casualties caused by them (approximately 50 per cent of Australian battle casualties), led to the employment of field engineer splinter teams with infantry and armour sub-units on operations.'

That's why engineers might find themselves walking ahead of APCs, checking for devices or clearing paths through the bush for infantry. Or they might be patrolling with infantry without a specific task but in the knowledge that an engineering challenge could be encountered. And, as Brigadier Greville

SPLINTER TEAMS AND MINI-TEAMS

noted, there was another very valid reason for the use of mini-teams and splinter teams: a lot of soldiers had to be trained up very quickly, and the only way to do that was on the job.

Thus, experience and know-how was passed from Sapper to Sapper, in the field, under combat conditions. It started with the men of 3 Field Troop and continued throughout the war. The 'rookies' would become experienced, passing on their skills to whoever was assigned to them in the next intake, newly arrived from Australia. This kept the experience and knowledge where it was needed most: in the theatre of operations.

There was never a time, after 3 Field Troop had invented the role of the Tunnel Rat, that all the acquired knowledge went back to Australia at the end of a tour of duty. In mini-teams, splinter teams and larger CET or ready-reaction teams, the word was spread and the culture passed on as the ever-adaptable Sapper found a whole new set of challenges against which to test his skill and ingenuity.

The splinter and mini-team system is still in use today wherever Sappers are engaged in warfare. Back in Vietnam, an added challenge for the Sappers was that they were often attached to infantry and armour platoons, so had to fall in with their cultures and methods. For some Sappers, that was as steep a learning curve and as alien an environment as going down into tunnels and bunkers.

5. ON PATROL

With all these changes, the life of a Sapper in Vietnam was an odd one, even for men whose job, by its very nature, required them to adapt and improvise at every turn. Attached in small teams to other units for the duration of an operation out bush for four to six weeks, instead of working together as a whole troop or squadron, they were expected to live and fight like infantry—or APC or tank crew, if they were with them—and still carry out their field engineering tasks like mine and booby-trap clearance and demolition duties. They often didn't know where they'd be going, who they'd be with or the nature of their mission until the day before they went out on patrol.

'You discovered which unit you'd be attached to, and where you were going from the Operations Board in the troop office,' writes Jim Marett. 'You'd wander in one morning to collect your mail and find out from the board that next day you're going out on ops with the grunts. The troop staff sergeant made the decisions and allocated the Sappers to the various units we were supporting.'

The Sapper's first task was to check and pack his gear, which included rifle and pistol, ammunition, explosives, detonation cord, detonators, maps, letter-writing gear and enough food and water for the first three to five days, depending on frequency of resupply by chopper. Some also packed 'Jack rations', such as cans of sardines or crabmeat, bottles of hot sauce, jars of Vegemite and packs of crackers. 'They were called "Jack rations" because most Sappers tried to extend the pleasure by keeping them to themselves rather than sharing them with mates—a "fuck you, Jack" attitude,' explains Jim.

Splinter teams would be attached to a company or platoon of infantry for the duration of an operation. The two-man team would consist of a more experienced Sapper and a number two, who had usually been in country less than six months.

'The operation would usually begin with a helicopter airlift into the area of operations (AO)—an event guaranteed to get the adrenaline going and the hairs sticking up on the back of your neck,' recalls Jim. 'A mass helicopter airlift of 12 or so choppers, filled with fully armed fighting men, is a sight to behold, especially when viewed from mid-air and when you are a part of it. The swoop into the landing zone (LZ) would climax the thrill.'

Once on the ground, up to six weeks of patrolling and ambushing would begin. In simple terms, the task was to follow signs of the enemy to their base camps, which would then be searched and destroyed. Sappers cleared the camps of mines and booby traps, searched bunkers for weapons and documents, then blew them up. After that, the Sappers also did their share of picket duty and manned the ambushes that would be set each night.

'Patrolling the jungle was done in the tested and proven Australian way—making absolutely minimal noise and leaving as little trace of your presence as possible,' recalls Jim. 'Communication was through an array of elaborate hand signals and an occasional whisper. Except for in the depths of a firefight, you never spoke above a whisper for the entire operation.'

Firefights were often in or around the enemy's bunker systems or a result of the ambushes set each night. The infantry were well trained in how to react to the start of a firefight so as to gain the initiative in the first crucial moments of contact with the enemy.

'In these first few seconds, one side or the other reached a point where it had the edge in the battle,' explains Jim. 'For the Australians, this process involved reacting without question and doing something totally opposite to what your mind and body would normally demand. You turned to face the enemy fire and moved aggressively towards them, firing as you went, then you quickly dropped to the ground to take cover. At the same time, the patrol's machine-gunners would head for the

ON PATROL

high ground to gain maximum effect from the weapon and give maximum cover to the patrol.'

As many engineers interviewed for this book would confirm, Sappers weren't anywhere nearly as well-trained in these tactics and often found themselves to be the only men standing, usually ten or 15 yards behind the rest of the patrol, who had disappeared from sight having advanced on the enemy then taken cover, all in a matter of seconds.

'Many Sappers felt a few days' training in infantry tactics would have solved the problem quickly and not left so many of them standing out like shags on a rock in their first contacts,' says Jim, who adds that, apart from the occasional firefight, the daily routine varied little. After a day of patrolling, a suitable site would be picked to harbour up for the night—preferably, a spot that was both a potential ambush site and provided reasonable sleeping positions for the men. They would spread out to form a circle around the harbour position, with machine guns positioned at its four 'corners'. The headquarters group would set up in the middle of the circle.

'You slept on the ground, with a groundsheet beneath you, but usually no "hootchie" tent above you, as they made noise in the rain and shone in the moonlight,' explains Jim. 'You cleared your personal sleeping area of scrub, hoping you didn't disturb an ants' nest or other bugs. Depending on the security situation, you sometimes dug a "shell scrape" to sleep in. This was disturbingly shaped like a grave, but only about 12 inches deep. It protected you from shrapnel, but not from a direct hit.

'If it rained, you simply got wet. Most men quickly discovered that by protecting their faces from direct raindrops, they could sleep through the rain. The evening meal was taken and then the patrol "stood to" before last light. "Stand to" was a process of silently listening and looking out from the perimeter during the time the patrol was traditionally most vulnerable to attack. After "stand to", everyone hit the sack—except for those first rostered to man the machine guns. In groups of two, everyone had a shift on the machine gun during the night, usually for two hours.'

Before first light, the group 'stood to' again, then shaved, squatted over a scrape-hole for toilet activities, ate breakfast and cleaned their teeth, before moving off on patrol. They couldn't have a decent wash for the six-week duration of the operation. No deodorant was used—the scent would easily be picked up by enemy forces—you didn't wash your hair and nobody wore underwear because the chafing it caused would cripple you in a couple of days.

'Bodies took on a unique mushroom-like smell after a few weeks—an aroma much like the moist earth around them,' recalls Jim. 'Often they didn't take their boots off for days, and when they eventually did, the men were frequently amazed at how their feet were in such good nick, considering the pounding they'd taken and the fact that they'd been wet and dry several times every day.'

Helicopters resupplied the patrols with food—and water, if they couldn't find drinkable water in the streams—every three to five days. Sometimes, depending on the season and the men's activities, their clothes would be so ripped or rotten that they'd have to be replaced with fresh ones flown in from Nui Dat.

'The feeling of putting on those fresh greens was pure heaven,' says Jim, who also recalls the surprisingly casual attitude many Sappers took to their equipment. 'Weapons were meant to be cleaned and lightly oiled every day. Sappers carried two guns, a 9-mm pistol used in searching tunnels and bunkers, plus either an SLR semi-automatic rifle or an Armalite fully automatic rifle. Ironically, many engineers were a bit slack about cleaning their weapons, at least until they had one fail in contact, which tended to focus them again on the importance of regular maintenance.'

However, Sappers were in their element when bunker or tunnel systems were found, which they frequently were. The bunker systems might be as few as three or four bunkers or a complex of 30 or more connected via tunnels or trenches. Sometimes the bunkers were defended, sometimes they were abandoned, often quite recently, with warm food and valuable documents left behind in the haste to escape. If the enemy defended the bunkers, air support could be called in, ranging from jets to drop bombs to helicopter gunships with rockets, grenades and machine guns. Artillery and mortar fire could be directed from the nearest fire support base (FSB), but because of the greater margin of error, attacking troops would have to pull back to a safe distance, which could give the enemy the chance to slip away. If tanks were nearby, they could also be called in; they were particularly effective in bunker assaults, with their massive firepower and intimidating size and noise. 'Once these boys entered the fray, the VC quickly and invariably decided it was time to *"ditty mau"*—to disappear or go away,' says Jim Marett.

After the fighting was over, an initial sweep was conducted through the bunker system. As they walked through the system, still on full alert, infantrymen usually threw grenades into each of the bunkers. 'This was great for the Sappers, because apart from flushing out any enemy, it also got rid of the bats and spiders they'd otherwise confront when it came time to search the bunkers,' observes Jim.

After the initial sweep, the infantry would form a defensive position around the perimeter of the bunker system. The company or platoon headquarters would set up in the middle of the position and the Sappers went to work. Their first task was to establish how many bunkers there were, and to call for a helicopter delivery of enough explosives to destroy them if there were more than just two or three. These would be dropped in if no suitable landing zone was nearby, sending infantrymen running for cover as the C4 explosives tumbled out of a hovering helicopter towards them. The Sappers knew it was perfectly safe, as long as no detonators were included in the same drop.

In addition to booby traps, for which Sappers were on permanent alert, bunkers held other potential surprises like bats, snakes and spiders. 'One species was a bit like Australia's daddy long legs, but with a much bigger body,' says Jim Marett. 'Sometimes there were so many that they covered the entire inside roof like a carpet, and as you moved, they moved in a wave, giving off a chilling rustling sound. The occasional pig or porcupine or armadillo might also be encountered, which usually resulted in a one-sided shooting match and a desperate dash for freedom by the animal and you.'

Once searched and cleared of any documents and weapons, the Tunnel Rats would set the bunker up with demolition charges, all linked by det-cord, that would destroy the entire system from one ignition point. Slots were dug into each interior wall, midway between the roof and the floor. A slab of C4 was placed in each of these slots, and on each end of the main beam supporting the roof, to cause maximum damage to the structure.

When all was ready, the infantry would move off and the Sappers would light the fuse and follow the rest of the patrol. The length of fuse determined the time before the explosion went off; depending on the terrain, which governed how long it would take to get away from the area, the fuse would usually burn for five to ten minutes. 'Waiting for the big bang was always an exciting and anxious time,' says Jim. 'And it was incredibly satisfying to watch the dirt, logs and debris hurled into the sky.' It was not unusual for a splinter team to search and blow up over 100 enemy bunkers in an operation.

Sappers were also called forward whenever the patrol encountered a landmine. 'This was without doubt the most harrowing aspect of the job,' says Jim. 'The enemy would often plant more than one mine, hoping to catch the men they knew would rush to the aid of those wounded by the first mine.'

A SAPPERS' WAR

If a mine had been triggered, Sappers had to clear safe paths to the wounded and dead so that they could be cared for or evacuated. Surrounded by the injured, who were often in terrible pain, and with the infantrymen understandably desperate to help their mates, a balance had to be found between rushing in, and ensuring that any other enemy mines were found and marked before they did further damage.

The success of a patrol was usually measured in the number of enemy killed, captured or wounded, and in the number of enemy bunkers destroyed. But eventually the operation would come to a close and it would be time to head back to the Nui Dat base camp. 'Choppers usually took us back,' says Jim. 'And what a glorious ride it was, knowing you were safe and that a shower, fresh greens and cold beers were just minutes away.'

6. THE MORE THINGS CHANGE . . .

Being deployed in splinter teams didn't mean Sappers were excused from their tunnelling duties. Far from it; they were expected to head underground wherever and whenever their infantry troop discovered a new entry. It was as part of a splinter team that Sapper Trevor Shelley had the kind of encounter all Tunnel Rats secretly dreaded—confronting a live enemy who wasn't in the mood to surrender.

'When 3 Troop was very close to leaving for Australia, I found myself as a number two with Doug Sanderson in a splinter team. We were in the middle of a night march, tied together with toggle ropes and with phosphorescent bark stuck in my webbing at the shoulder. It was very, very dark. We were with 5 RAR and about to pull a cordon around Duc My village. All of a sudden there was a halt—toggle ropes went taut and a quiet murmur moved down the lines. A 5 RAR rifleman had fallen down a well in the dark, so we all waited until he was dragged out. It was miraculous that he did not hit any cross members on the way down but he was extracted quickly and, soaking wet, rejoined the somewhat eerie trek through heavy jungle. There was a plan to this, as I realised just before dawn. We—5 RAR—had put a cordon around a village, and from the initial gunfire, some in our direction, some residents were obviously angry at being disturbed so early in the morning.'

Doug Sanderson takes up the story: 'We arrived at our planned position in the cordon around Duc My before dawn. Then a light aircraft flew overhead, with a speaker system telling the occupants to stay in their huts as the village was surrounded. There were some VC inside the cordon who tried to make a break for it. This resulted in one dead and a number injured. Once things had settled down, we moved in to search the village and question the inhabitants. It was our job as Sappers to do the hut and bunker or tunnel searches.

'The little huts were only small one-room structures that were used for every purpose—sleeping, cooking and so on. Most had a bunker for shelter in the event of attack. In one of the first huts we went into, we found a detonating device. Our next move was to look in the bunker. We removed our webbing and armed ourselves with our 9-mm Browning handguns.'

After careful inspection of the entry, Doug jumped down into the bunker. He heard some noises and asked Trevor if he had heard anything. He wasn't sure. 'In we went,' Trevor says, 'with Doug in the lead—he was probably more worried that I was coming up his backside with a 9-mm handgun than what was ahead.' Doug and Trevor carefully made their way into the bunker to discover that, in fact, it was a small tunnel, not one cavern. After about three metres, the tunnel led to a T-intersection.

'Very gingerly, we looked around the corners,' continues Doug. 'Trev was to one side and I was to the other. We could not see to the end of either side as the tunnels curved away—this would have been done so there was no line of sight for shrapnel or bullets in an attack on the system. Of course, we were now faced with the dilemma: which way should we go first? Or should we even split up? I decided to get Trevor to cover me from the intersection as I recce'd to the right; he was also to watch to the left, in case anyone was down that side who could come behind, trapping us.

'On working my way forward, I heard more noises. I then went back to the intersection, which was now out of sight due to the curve. I told Trevor what had happened and said I would monitor the intersection, and I asked him to check out the left arm. He was not to go too far out of my sight. Also, I was now certain that there was at least one person in the system, but I didn't know its extent. When Trev returned he reported that it only went about 20 feet and it was clear. Now that I knew that the left branch was clear, I decided to explore the right side. I asked Trev to follow me. And, yes, the thought did concern me that he would be following with a loaded gun.

'As we moved along I heard more noises, and as we continued along around the bend I saw a person about three metres in front of me. He was standing on a shelf going up to an exit shaft. I could only see from his thighs down, so I had no idea whether he was armed with a rifle or grenade ready to do the suicide thing. Not wanting to go any closer yet, I called out for him to come down and show himself. No response.'

Doug sent Trevor back to get an interpreter, but the South Vietnamese soldier would only jump down into the entrance and shout. Doug fired a warning shot but that had little effect. 'At first I thought my Browning had misfired, so I fired again and once again there was just a pop,' he recalls. 'I then realised that the earth of the tunnel was acting as a silencer. My next thought was that I might have to shoot him; this gave me a moral dilemma. What if I killed him and he was unarmed? On the other hand, if I shot him and he had a grenade with the pin pulled, we would all be gone.'

Doug decided that tear gas was the best option, so he and Trevor pulled back, put on their gas masks, grabbed a CS grenade, went back in and released it. 'Doug lobbed a grenade under him as we could only see his legs,' says Trevor. 'We retreated and sat on top of the bunker for a while, and he seemed not to like it much as there was a fair amount of noise coming from inside the bunker. We then pulled back to allow the gas to do its thing and covered the entrance, waiting for him to come running out shooting.'

Doug and Trev waited a little longer for the gas to clear—when the gas is first released it fills the area with smoke, making it impossible to see. They went back in and found the man lying unconscious at the bottom of the exit shaft;

he was not breathing. 'He had decided to take the gas and not run out, confirming my thought that he was prepared to commit suicide,' says Doug. 'I then checked him out for any form of booby trap, declared him and the area safe and asked the infantry guys to extricate his body through the exit shaft.'

After some time the man was dragged out, and the 5 RAR medic took over and revived him. It turned out that he had received leg wounds during the morning cordon actions, which had forced him to seek refuge in the bunker. 'The elderly lady who owned the house was distraught—she was in fear of her life due to repercussions from the Vietcong because he was found there,' says Trevor. 'I managed to salvage his hammock, which I used for the next ten years, as well as his cigarette lighter and his knife, which I still have today.' According to many reports, this event gave the Vietnamese an insight into the Australian character—the Sappers had taken the more humane of two options. 'In our webbing we also had M26 hand grenades,' says Doug. 'The soft option was the tear gas.'

The prisoner had rounds on him and would have had a rifle, but one of his companions must have grabbed it to prevent it from being captured. Doug was recommended by Sandy MacGregor for a bravery award, but because of the quota system on awards it was never given. 'I am not at all concerned about this,' says Doug. 'Trevor Shelley was with me, and he would deserve the award also. That's life!'

Sandy didn't discover until long after the war was over that, at that time, Doug Sanderson had suffered from claustrophobia. Had he known that, Sandy says, he would never have sent him into tunnels but Doug never said anything. 'It was not until November 2009 that I asked Doug how he did it,' says Sandy. 'Doug said, "I told myself that I had to do it—it was my job. I then pictured, imagined and told myself that I was crawling on the surface of the earth with plenty of air to breathe." What a wonderful example of mental toughness achieved through focused concentration and visualisation.'

Going down a tunnel when you don't know who or what is around the next corner—and, by design, there are a lot of corners—is terrifying, even for those who don't have a problem with confined spaces. You have to wonder what it was like for the Vietcong hiding in the tunnels during a bombing raid. Trevor Shelley experienced this first-hand during another splinter team tunnel search.

'We were moving along a quite straight length of tunnel, which was pretty long so the light illuminated the whole length,' recalls Trevor. 'It had the usual domed roof and was a hands-and-knees job. Without warning, the whole length in front of us started to buckle and there was a rumbling noise. Large clods dropped out of the roof and walls, and just as fast as it began it stopped, leaving us quite shaken up and in a tunnel full of dust. Needless to say, we got out of there. To this day I have no idea what caused it, except that somebody in the platoon mentioned that there was a B-52 strike nearby.'

Tunnels were an ever-present feature of the war, as Allan

THE MORE THINGS CHANGE...

'Blue' Rantall discovered when he joined 1 Troop, 1 Field Squadron in October 1967: 'I had only been in the army long enough to have done Corps training at the School of Military Engineering and a couple of months out bush at Tully working on the range at field engineering tasks, the Officer in Command there was John Wertheimer and the second in command was "Flex" Fittock. I arrived in Nui Dat on the 10th October 1967, I was 22 years of age.

'When I first got there the Squadron was building the new village of Ap Suoi Nghe, a couple of "clicks" out from the end of Luscombe airfield. At the same time the "House of Horrors" [a booby trap museum] was being put together under the direction of Sgt Brett Nolan and we were their first pupils in learning the "tricks of the trade" for sappers in Vietnam. My first operation was with C Company 2 RAR and we were delivered to the start point by APC. I was so scared that I tippee-toed around the place for the first day or two until I started to relax. I started off as a brand new fresh-faced number two to Barry O'Rourke. We had not been too long into the operation when the Company came upon a big tunnel complex, quite a bit of fighting went on to clear the system and then we were given instructions to blow it up.'

The task was much too big for just the splinter team so Barry and Allan called in the combat team that was waiting at the FSB. They faced a pretty gruesome task, as there were dead bodies scattered throughout the complex. 'In one nook I came across a decomposed corpse that had obviously been there for some time,' Allan writes. 'Between the splinter and combat teams it took us about half a day to rig it all up for demolition, it was a very big complex.

'A few days after this another tunnel complex was discovered but this time there was a bit of a snag. The infantry tossed in hand grenades at first but when we started to go down we discovered that a lot of the grenades hadn't gone off. Barry was a very good number one and he was not going to take any unnecessary risks so he got a bit stroppy and insisted that we would neutralise the enemy with our very own home made grenades. So, that night, we worked into the wee hours of the morning and using slabs of American TNT (not C4) we rigged up hand grenades using detonators, det cord and very short lengths of fuse cord so the VC could not neutralise them in time.

'Barry and I were having a well earned brew while waiting to move down into the tunnels when all of a sudden a VC popped his head out of a bunker and started firing on us. We were the closest to him so the Company Sergeant Major ordered us sappers to do a frontal charge and take the position. All that was scary enough but what really pissed me off was the fact that in our mad haste to do the "gung-ho" stuff I knocked over my brew and being in the middle of the dry season I didn't have enough water to make another one.'

The next time Allan had to venture into a tunnel almost proved his undoing. 'It was my turn to go down first with Barry following. I had only gone in about ten or twelve feet when I felt I had knocked something hard and metallic, turned out it was a CS gas canister—about the size of a kerosene tin. It had been lying dormant and my knock must have been enough to set it going. I was in a lot of distress and choking to death very quickly, fortunately Barry had enough wits to grab hold of me and pull me out by the ankles.

'It must have been a Herculean effort from him as he too was affected by the gas. When we got out I was very sick for an hour or so and did a lot of vomiting while trying to get my lungs working again. As to how it got there all I can think is that the canister must have been left over from a U.S. operation or the VC had put it there themselves. But whichever way, they nearly got me—thanks Barry!'

Allan's rescue demonstrated the strengths of the two-man splinter teams, which was anathema to traditional army engineers but a major step forward for Sappers. For Jim Marett, working closely with other units had both advantages and drawbacks, with Sappers always on the lookout for the ultimate swan. 'Occasionally, if you were lucky you'd get attached to the APCs for an operation,' he says. 'Living in close habitation with the "tankies" generated regular disputes over sleeping space and the sharing of Jack rations, but at least you didn't have to walk through the scrub. If you were really lucky, you'd spend an operation on ready reaction at the FSB. This was a total "swan", with regular showers, mail and meals. Mostly, though, we ended up attached to one of the infantry battalions, walking with the grunts.

'In addition to our mine, booby-trap, tunnel and bunker searching, and our demolition duties, working with the grunts allowed us to act as honorary infantrymen. This had more to do with filling the picket duty roster on the gun each night than with any admiration of our soldiering skills. Had the grunts been aware that our total infantry training consisted of the real basics at rookie camp and the crash course at Canungra, they may not have welcomed us so openly.'

Jim was attached to A Company of 5 RAR for Operation Kings Cross, which ran from 31 October to 12 December 1969. 'It was an extremely arduous operation where we found lots of bunker systems—all of which had to be searched and then destroyed,' he says. 'One we came across was very well camouflaged. Fresh grass clippings had been spread across the trails leading to the entrances in a bid to conceal them. The bunkers contained heaps of medical supplies and equipment, mostly of French origin. As we made the initial sweep through the system with the infantrymen, grenades were thrown into each of the bunkers. Unfortunately, one of these grenades failed to explode.'

Company commander Major Reg Sutton made the decision that they would harbour up within the enemy's camp to see if the Vietcong would return for their valuable medical supplies. This postponed the task of blowing up the bunkers, but Jim was given the job of getting rid of the unexploded grenade.

'Not wanting to move his men or to make undue noise, Major Sutton said I should make the explosion as minimal as possible,' Jim recalls. 'He also wanted my assurance that he and his men would be safe from the explosion in their present positions. Major Sutton and the rest of the HQ group were situated right alongside the bunker containing the grenade.

THE MORE THINGS CHANGE . . .

Calling on my entire ten weeks of corps training and subsequent in-country education, I gave the thumbs-up and headed into the bunker. The last thing I wanted to do was touch this grenade in case I loosened something that would set it off. I figured that half a stick of C4, gently placed beside the grenade, would do the trick. I came out of the bunker to make up the charge, at the same time suggesting to Major Sutton that everyone could stay where they were, but to be safe they should lie flat on the ground.

'Returning to the bunker, I put the charge in place, lit the fuse and exited with the traditional cry of "fire in the hole". Perhaps as a show of confidence I took my position, flat out on the ground, right beside Major Sutton. Waiting for a demolition charge to go off is always a weird mix of tension and excitement. With my high-ranking superiors gathered about me, this wait was particularly tense.

'Suddenly the earth shook. The ground rumbled and tossed beneath us. I caught a glimpse of Major Sutton and the others—their bodies were flapping like fish freshly landed on a river bank. This was not good. Leaves and twigs rained down on us and a layer of dust lifted three feet off the ground, briefly forming a cloud before settling on us. It was a sorry scene and I was quickly becoming the centre of attention. I prepared myself for the worst as Major Sutton rose, brushing the debris from his greens. He threw me a cruel stare that could have cut through steel, and said, "Back to the drawing board, Sapper."'

The company waited in ambush for a couple of days, but 'Charlie' didn't come back so Jim eventually set up the charges and blew the bunkers, using a particularly long fuse to enable everyone to get a very safe distance away before the big bang. 'I must have been forgiven by A Company because they had me back with them on Operation Bondi,' laughs Jim. 'They even took me with them on leave to Vung Tau... but that's another story.'

7. FOOT SOLDIERS

The Sapper, whatever his engineering skills, is so close to being an infantryman that sometimes it's hard to tell the difference. Even if he's a bulldozer driver, probably capable of driving a tank, his fighting DNA is closer to infantry than to a tankie's. When a Sapper is defusing mines, clearing tunnels and 'delousing' booby traps, he is 100 per cent engineer. But when those tasks are done, he can share infantry duties like any other grunt. This is just as well, because once the splinter teams became established as standard operating procedure, the Sapper often spent more time with the infantry than with his own kind.

In all, six Sappers were awarded the highly prestigious Military Medal in Vietnam, five of them for their work in support of the infantry. There would probably have been a lot more but for the army's illogical rationing of honours. The exploits of Neil Innes and Phil Baxter have been described elsewhere in detail, but the heroism of Lionel Rendalls, Ray Ryan and Gary Miller is typical of the vital work Sappers did with the infantry.

On 17 October 1966, Corporal Rendalls was the leader of a splinter team attached to an infantry company that suffered seven casualties from booby traps as they were moving into a village. Corporal Rendalls moved to the head of the company and cleared forward, covered by only one scout to his rear. For the next two hours he worked alone, finding and delousing 17 booby traps, allowing the infantry to move on without further casualties.

On 10 March 1969, Sapper Ryan was the team leader of an engineer splinter team attached to D Company of 5 RAR. In the middle of the night, soldiers entered a minefield and suffered several casualties from mines and small-arms fire. Sapper Ryan cleared paths to the wounded, working single-handed in the darkness for four hours, using a bayonet and a torch. After he had reached the last of the wounded and was leaving the minefield, he detonated a mine and was badly injured himself.

On 6 March 1970, Sapper Miller was the leader of a two-man splinter team supporting a rifle company of 8 RAR when an enemy mine was triggered. Sapper Miller immediately started clearing safe lanes by hand to allow access to the casualties. He then cleared a pick-up area so that dust-off helicopters could winch the wounded out. Just over a week later, Sapper

FOOT SOLDIERS

Miller was again supporting an infantry force when a mine was detonated, injuring four soldiers. Once again, he cleared safe lanes to the wounded; knowing the Vietcong often laid secondary mines a little away from their main clusters, he searched until he found another anti-personnel mine 60 metres from the initial explosion. It was in an area through which troops would have passed in order to avoid the marked minefield.

It's in the nature of soldiers to celebrate their differences, but Sappers and infantry grunts had mutual respect, if only for the simple reason that they were in the business of keeping each other alive. Engineers would painstakingly inch forward, detecting and neutralising landmines so that the infantry behind them had safe passage through a potentially lethal patch of ground. In return, the infantrymen would cover the engineers' backs when they had to focus all their attention on the task in hand, saving none for the possible approach of a deadly enemy.

One difference worth noting, though, is that while the engineers in splinter teams would cover the same ground as their infantry mates, they did so under the additional burden of all their engineering equipment, which added an extra ten or 15 kilograms to the basic load hefted by other foot soldiers. And there was a lot of time between engineering tasks when Sappers were expected to do basic soldiering. Trevor Shelley had a typical induction into the ever-changing world of the Sapper in Vietnam.

'My nineteenth birthday had passed a couple of days earlier

when I began a rather circuitous route to eventually arrive at 1 Field Squadron in Nui Dat,' he recalls. 'We left on a Qantas charter out of Richmond on 25 April 1966 bound for Manila, where we were shuttled onto an Air France charter into Saigon—but of course a lot of our gear was not offloaded in Saigon and ended up in Paris. Three weeks later it all came back.

'Hitting Saigon, it was a pretty quick transit onto one of those tiny Hercules planes with two engines—it leaked like buggery in a rain storm,' says Trevor, who had probably enjoyed the delights of a Caribou transport plane, used for small loads and short runways. 'Then we were off to sleep in the sand hills of Vung Tau.' There were very few Australians in Vung Tau at that stage, as the HMAS *Sydney*, carrying 5 RAR, wouldn't arrive until 6 June. So Trevor settled into life with 10 Troop, 17 Construction Squadron.

'The only big action event I remember during my stay with 17 Construction was when a bloke from the workshop went loopy and was running around the sand hills with a loaded Owen gun,' recalls Trevor. 'He eventually ran into the sea with the threat that he was going to swim to Australia and then gave up. I eventually travelled to Nui Dat in early June, I think, and cozied into the 1 Troop lines. As a Sapper, I wasn't very aware of the whole situation of tactics and operations. We just tagged along with the companies in splinter teams. The only difference between us and the normal grunt was that a) we carried more gear, b) we were often tinged with a nice shade of red from the work we embarked upon and c) we were supposed to be more intelligent.

'This operation was a big one for 5 RAR. It started when we were winched into the top of "Wolverton Mountain", with chainsaws an unwelcome addition to the rest of our gear. And to work we went, cutting down trees to build a big helipad, while a company of infantry protected us. It was hard work, and the coldest nights I have ever experienced in Vietnam.'

With the task accomplished and the chainsaws flown out again, Trevor and the other Sappers joined the rest of the battalion, which was moving along the top of the range. He recalls it was hard country and hard walking. There were lots of niggling little skirmishes and many rocky bunkers and pagodas to clear.

'When possible, bags of CS tear gas in a powder form were flown into us to neutralise the bunkers and tunnels,' says Trevor. 'By firing a slab of explosive under the bag, it would impregnate the walls and crevices, making it damned unpleasant to be in for a couple of months. I thought then that I had this practice down pat, until one afternoon when, like an amateur golfer, I did not allow for the wind. On emerging from my safe spot I glanced up to see the wind blowing the residual powder left over from the blast. There was a large number of Aussies dug in but they all had tears streaming from their eyes and were somewhat miffed. It took a few days before we were mates again.'

Trevor recalls that the platoon members had become very jaded with the operation, as they had been out for several weeks without any respite. They had gone into a harbour position fairly early in the afternoon, for a change, and those who weren't on sentry duty or at listening posts were setting up for the night, or just smoking and 'spacing out'.

'I saw a couple of choppers come in and out of a temporary helipad but I didn't take that much notice as we were in a decent position and had need for resupply,' says Trevor. 'About an hour later, when everybody had settled down and were just in a dull fuzz of fatigue and boredom, we noticed a small group of senior NCOs making their way slowly through the area and stopping at each small group for a short time. As they travelled through the groups a soft murmur sprang up.

'We were still in the dark as they approached where we were dug in and hoochied up, and it was then that I noticed that they were carrying a couple of large stoneware flagons. A gruff "Get your mugs out!" was conveyed by the company sergeant major, which we did ... and then I smelt it! Was this a hallucination or was that smell really *rum*?'

A generous tot of navy rum was issued; it was of great quality, or so it seemed to Trevor at the time. 'Somebody in a senior position had the common sense to do something simple to raise flagging spirits,' says Trevor, 'and it worked. Now I understand the tot of rum issued in trenches of the Western Front during the First World War before they went over the top.'

The trenches also came to mind for Allan 'Blue' Rantall in February 1968, when a large element of the Task Force—including all of 1 Field Squadron—moved out to FSB Andersen and took over the position from the Americans. An FSB was, to all intents and purposes, a temporary fort protected by earthwork walls and housing elements of infantry, armour and artillery, often in or near an area where there was a lot of enemy activity and too far from the main base for protection. The infantry and armour—usually APCs—would go out on patrol from the FSB and would be able to call in artillery and mortar fire on any enemy positions they encountered. The main body of troops would return to the FSB at night, although some might remain outside to set ambushes for any Vietcong or NVA forces that might attack.

Typically, the Sappers' first job would be to build or reinforce the protective walls, often by bulldozing earth, laying barbed wire and setting Claymore mines. Claymores are anti-personnel mines designed to blow out towards advancing troops; they are usually fired by wire linked to 'clickers' held by troops in the defensive line, or by trip-wires. As Allan recalls, there were always ways of making the best of the terrain. 'It was shades of the First World War for us when, with the help of a dozer, we built one very big and deep bunker that was able to house the whole of the troop,' says Allan. 'Other than the fact that it felt like one huge grave, it was quite cosy.'

One of the Sappers' tasks was to provide a night time standing patrol about 1.5 kilometres out from the FSB, overlooking the village of Trang Bom. 'The standing patrol consisted of about four of us Sappers moving out at dusk each night and keeping a watch from beneath a tree,' he recalls. 'The VC were very active in the area and attacked the village on an almost nightly basis. We would watch all the firing going on and houses being burned, then the next day we'd join clearing patrols to detect for mines and booby traps before the infantry went in.'

On 18 February it was 3 Troop's turn to provide the standing patrol, but the Vietcong launched a mortar attack on the tree at the same time as they mortared the FSB. Lance-Corporal John Garrett and Sappers Allan Pattison and David Steen were all killed.

'After the attack, which was in the early hours of the morning, I left our shell scrape to check on the surrounding area for casualties,' remembers Allan. 'I came across an artillery digger who had taken a direct mortar hit. It turned out that he was an old schoolmate of mine from Alexandra. We had actually flown into country on the same flight. James Menz was his name, and he had only just got married two weeks previously while on R&R. So it was a very tragic and upsetting time for all of us that night.'

Australian Sappers didn't only work with their own infantry. Rob Woolley and his number two, Doug Myers, were out on patrol for several days as a splinter team with New Zealand Victor Company, attached to 6 RAR, when the forward scout spotted a weapons cache. Rob and Doug searched the area, and found a stack of weapons abandoned by the enemy, who must have been caught by surprise. The guns included two Spandau machine guns that had never been used and still had packing grease around them. In a strange coincidence, during the Second World War, Rob's Dad was shot and wounded at El Alamein in late 1942 by a tracer bullet fired from a Spandau. The bullet went into his chest and out through his neck but he survived and was fit enough to take part in the Finchhafen landing in New Guinea in September 1943.

Coincidence and luck played a big part in surviving Vietnam. Corporal R.J. 'Stiffy' Carroll recalls his very first operation there—a booby-trap clearance with Ross Thorburn and Cul Hart—where his number seemed to be up. 'It was memorable for the fact that, as I was disarming a Chi com grenade rigged on a trip-wire, a young grunt kicked the wire,' says Stiffy. 'I had my hand around the grenade at the time and was about to cut the wire. Somebody up there must like old Sappers because it failed to explode.'

Stiffy remembers another incident when a second of his nine lives was used up. 'Four of us—Sandy Saunders, Jim Garner, Lance Guthrie and me—went out on a routine sweep with a Defence and Employment platoon, looking for mortar base plates. It was supposed to be a cakewalk. We rode on APCs to our rendezvous, where we were told by the young lieutenant to make ourselves comfortable and have a feed as there would be nothing to do unless someone got into trouble.

'I took off my small pack, which contained four slabs of CE explosives and a box of detonators, and put it on the ground while I opened a can of ham and lima beans. In the background we could hear the lieutenant and the radio operator calling coordinates over the radio. Sandy, who was a well-known cynic, gave me a nudge and, in a horse whisper, said, "I hope they taught him properly—he doesn't look old enough to have left school."'

The quartet continued eating and having a bit of a yarn

when they were interrupted by a series of popping sounds from a distance. 'Our 105s,' said Sandy Saunders, answering Jim Garner's enquiring look by confirming it was their own artillery firing at them. 'Then came a sound that was unforgettable,' recalls Stiffy. 'It was a ripping sound, like the noise made by a truck's tyres on a hot bitumen road. As it grew louder, Sandy dropped his can of beans and yelled, "It's us!"

'He grabbed Jim and threw both of them into a small depression against the side of the track. I followed suit, throwing Lance down and landing on top of him, putting my leg over his head and trying to burrow my own head into the tangle of limbs. The first round, when it exploded, must have been directly overhead. The concussion was like being whacked by the flat of a square-mouth shovel, and the noise was a physical assault. I have no idea how many shells exploded—probably only five or six—but lying there directly underneath them, they just seemed to go on and on. Then there was silence, or a sort of silence with a whistling ringing noise. The radio operator, who had been sitting next to us with his radio on his knees, had been hit in the back and was obviously dying.

'I was first up and went to his assistance. I whipped off my lucky shirt and covered his gaping wound while I cradled his head and screamed my head off for a medic, who thankfully was on the spot immediately. Sandy and Jim were looking dazed and Lance was groaning and writhing on his back on the ground. Sandy started to tell him to stop whingeing when we both noticed smoke coming from under him. We rolled him over, and there sticking out of his shoulder was a three-inch piece of shrapnel, hissing and smoking. When I went to retrieve my pack with its explosives, there was nothing but a white depression in the red soil. I knew I should never have put my dets in the same pack as my CE... nor my rations, for that matter.'

But that wasn't the end of the story for Stiffy. 'For the next month I had a terrible time—I couldn't seem to understand what people were saying to me. I heard them and pretended to understand what they were saying, but nothing made sense. It was as though my memory had gone AWOL. I began to avoid talking to my men and sent my lance-jack [lance-corporal] Ron Forsyth, to the operation planning meetings. One day, my good mate Troop Sergeant Blair Parsons pushed his face into mine and told me I was going to Vung Tau to see the headshrinker. "There's something bloody wrong with you, mate—you'll get killed," he said. "I've been blowing a bloody whistle right behind you for two minutes and you haven't raised an eyebrow."

'Twenty-four hours later, at 36 Evac Hospital, I was declared totally deaf in one ear and unable to hear in the other. I was downgraded from FE (field engineer) to HO—that is, home only. I was not even allowed to go back to Nui Dat to pick up my gear. Seventeen Construction Squadron was in Vung Tau, so I was told to report to them to await "RTA"—return to Australia.'

8. CORAL, BALMORAL AND BEYOND

Often Sappers weren't just working with infantry, they were fighting *as* infantry. And nowhere were they more directly involved in combat, defending their own and their comrades' lives, than during the battles for FSBs Coral and Balmoral.

After the Vietcong and NVA made inroads into Saigon in the 'mini-Tet' offensive of May 1968 but were driven off— the city's population failed to rise up and support them, as they had hoped—the Australian Task Force was redeployed to block their withdrawal from the capital, with two battalions establishing an FSB named Coral just east of Lai Khe in Binh Duong Province, 20 kilometres north of Bien Hoa. Allied commanders suspected their plans would meet some resistance, but that would turn out to be the understatement of the war. The resulting battles for Coral and

Balmoral would be Australia's costliest and longest engagement in Vietnam.

FSB Coral, would sit astride the main route used by the communist forces for attacks on Bien Hoa—the American Base—and Saigon itself. The next FSB, probing even further into 'bandit' country and named Balmoral, would be used for patrols and assaults on enemy positions. There was no way that North Vietnamese forces, who were present in large and well-organised numbers, despite their defeats in the Tet and mini-Tet offensives, were going to allow either of these intrusions into 'their' territory without a fight.

But first someone had to build these FSBs, and that, of course, meant Sappers. On 13 May, Sergeant Phil 'Jonah' Jones, Sapper Edwyn Stanton and a CET landed at Coral. They were there supporting an advance party that included infantry and elements of 102 Field Battery, 1 RAR's mortar platoon, and some New Zealand artillery. Their arrival did not go unnoticed, and it didn't take long for them to find themselves in action.

Edwyn Stanton, who had been a kangaroo shooter before the war, shot three NVA reconnaissance soldiers from the seat of his bulldozer while he was building bunds to protect the artillery and mortar units. However, there had been considerable confusion in the planning of the FSB, which led to delays, meaning it was not properly prepared by the time night fell. In the early hours of the next morning, NVA commanders, realising that Coral was never going to be more vulnerable, launched a ferocious attack, during which Sappers were deployed alongside infantry, purely as combat forces. It was a close-run thing. One artillery piece was briefly captured, before being retaken, and a mortar team was almost knocked out before the enemy were driven off.

'The main body of the Task Force arrived the next day, but one of the first jobs for 1 Field Squadron was to dig a mass grave using a bulldozer,' recalls George Hulse. 'Then we had to assist with the pick-up and burial of the NVA dead, showing them as much respect as possible.'

Allan 'Blue' Rantall was on the convoy that brought 1 RAR to FSB Coral. 'It's said that the convoy that went to FSB Coral was the biggest Australian convoy since the Second World War, and I have the distinction of being on the first APC into the location,' he recalls. 'But first we spent an overnight stop at Bien Hoa, where I had my first taste of spirits and pizza.

'A mighty strange place was Coral. At first we had to crawl around on our hands and knees so the Vietcong wouldn't pop us off, and at night they would creep up close to us and throw rocks so as to get us to return fire at them; then they'd pick us off by firing at the flashes. Someone in authority had the bright idea that we should mount a guard up a big tree that was beside our gun pits. That was a scary job . . . I reckon the song "Bad Moon Rising" was written about that bloody tree.

'Within days, the dozers had pushed back the tree line to create a bit of clear space between us and the enemy. We had been issued with a Starlight night-vision scope by this stage and were able to observe the enemy as they moved around the perimeter. As you looked through the scopes everything was a bright green—spooky stuff.

'Then I had a bit of luck. Somehow, John Wills and I from 2 Troop were selected to be choppered back to Nui Dat one morning to provide the mini-team that would escort the tanks up to FSB Coral, once again with a stopover for more spirits and pizza at Bien Hoa. Most likely, that was one of the longest trips the tanks ever made during the war, and John and I got to do it with them.'

The two mottos of the Sappers—'Everywhere' and 'We Make and We Break'—have rarely been better illustrated than during the construction and defence of FSB Coral. 'During the day we were deployed on mine clearance, field fortifications, water supply, road construction, helicopter landing zone development and land clearing to deny 7 NVA the use of covered approaches,' says George. 'At night we became a de facto infantry company. Several times we were attacked by NVA ground troops, which were supported by rockets, mortars and artillery.'

The next big attack came early on the morning of 16 May, when two battalions of North Vietnamese soldiers were launched at Coral. One RAR bore the brunt of the onslaught, but it was only under four hours of withering fire from small arms, artillery and mortars, helicopters and C-47 planes equipped with mini-guns and illumination flares that the North Vietnamese withdrew.

By the time dawn broke, the battle was all but over. Five Australians had been killed, and 34 North Vietnamese bodies were found. However, there was no definitive figure on how many enemy had been killed or wounded in these battles; whenever they could, they tried to remove their dead as they retreated.

Apart from the major battles, Australian troops would go out on patrol, looking for stragglers or any build-up of North Vietnamese forces. This, too, led to firefights, some of which were ferocious affairs, lasting for hours. It was during one of these that Private Richard Norden of 1 RAR rescued a wounded man under heavy fire, for which he was awarded the Distinguished Conduct Medal.

Once Coral had been established as a fully functioning and well-defended FSB, it was time for phase two, and about five kilometres north of Coral, FSB Balmoral was created on 24 May. Naturally, Sappers were there early in the piece when, once again, Sergeant Jonah Jones and his CET were sent forward. This time, they were assigned to a 3 RAR battle group, with the job of building defences and repairing them after the inevitable NVA attacks.

Among his other duties, George Hulse was tasked with setting up one of two Claymore minefields, which would be triggered by trip-wires. However, the most effective defence was simply barbed wire. 'The value of wire as an obstacle can't be overstated,' George says, recalling that attacks by the NVA were designed to overwhelm the Australians by sheer force of numbers. The first time they did so, when there was no wire out front, they almost achieved their aims. The next time, just one coil of wire was enough to slow them. In fact, in both major attacks on Balmoral, the NVA eschewed the cover of surrounding bush and attacked front-on, where there was less wire but, as they found to their dreadful cost, a greater concentration of firepower. It would turn out to be a bloody exercise for both sides.

The battles of Coral and Balmoral were the first time Australian troops had fought well-trained and disciplined NVA soldiers. By the time Australian forces were withdrawn from the FSBs in early June 1968, some 25 Australians and several hundred North Vietnamese and Vietcong fighters had lost their lives.

The Tet offensive and its aftermath were militarily disastrous for the communist forces but many observers identify this as the point at which the tide of popular opinion started to turn against the war, both in America and in Australia. The Vietcong and NVA forces had previously been portrayed in the Western media as badly organised, massively outgunned and pathetically undertrained. This was clearly no longer a sustainable view: the North's forces had lost badly but they had put up a tremendous fight and the perception that they were little more than a fanatical rabble whom the Allies would inevitably defeat suffered its first serious dent.

9. KNOW YOUR ENEMY

Set-piece battles, defending a position against an organised attack, as happened at Coral and Balmoral, were the exception rather than the rule for Australian forces in Phuoc Tuy. They engaged the enemy in jungle warfare, using stealth against stealth, preferring to try to outwit rather than outgun them. John Tick, then a young captain, recalls how a chilling encounter with an unseen enemy might have been disastrous but for smart soldiering.

'It was early 1971, during Operation Overlord, the purpose of which was to close with and destroy 3 NVA Regiment, which was known to have a base area on or near the Long Khanh-Phuoc Tuy border buffer zone,' says John. 'When 3 RAR eventually encountered the enemy and fought through their bunker position with tanks, a Sapper CET, protected by their old stalwarts the Pioneer Platoon, were instructed to find, map, search and demolish the enemy bunkers.'

Pioneers were, first and foremost, infantrymen. When required, however, they also performed some basic technical tasks such as demolitions and installing wire obstacles and overhead protection (OHP)—usually of logs, steel, sandbags and dirt for company headquarters.

'Our initial exploration of the bunker system revealed it to be about 1500 metres long by 1000 metres wide—a huge system,' says John. 'It was well dispersed, too, with more than 100 bunkers, each able to accommodate six to ten men. On the night of day three, the combined demolition team established a night harbour—a defensive position—just outside the bunker system. It was our habit never to occupy any night harbour a second time, for fear that the enemy would detect its location and mortar it during the second or subsequent occupations. Worse still, they might mine it while we were away doing our jobs.

'This particular night, we were all pretty buggered as we had run very low on rations and water; the demolition task had taken much longer than we had anticipated. Two Troop had a very good working relationship with the 3 RAR pioneers who were our protection and support group during the frequent demolition tasks we undertook. Tactical control of the group remained with the pioneer platoon commander, while we undertook our Sapper tasks. When not on demolition or other Sapper tasks, we contributed to the infantry's routine requirements for gun picket duty, clearing patrols, night radio watch and so forth—it was a good symbiotic relationship.

'The night passed as nights in Vietnam did, with odd bursts of machine-gun fire here and there and the occasional firing of artillery and mortars. At about 3 am one morning, Lance Corporal Roy Sojan, one of our Sappers, crawled into the platoon headquarters area and gently woke me up to come and listen at the gun pit where he had been doing picket with another Sapper. We crawled very quietly the 20 or so metres to the machine gun position and lay listening to the odd jungle noises outside our perimeter.

'Soon after arriving, we heard an intermittent *click, click*—a sound uncannily like someone snapping their fingers. My heart missed a beat or two—the source seemed human, possibly from an NVA recce team. We lay dead-still, not whispering, and listened as the clicking moved quietly around our perimeter, maybe ten metres or so out.' Roy asked John what they should do next. 'Absolutely nothing!' John said, not wishing any noises from unexpectedly roused infantry or Sappers to alert the enemy to their positions. John warned Roy to stay quiet but also to make sure he cautioned the picket replacements. 'The last thing we wanted were a few rocket-propelled grenades fired into us,' he explains.

'I crawled back to the platoon headquarters position but decided not to wake Peter Abigail, the platoon commander, or his sergeant or radio operator. I just lay there wide awake till just before dawn, and we went into our daily routine. The next day we continued searching, clearing and blasting, and that night we set up a new harbour some 300 metres away from the old one. The cavalry carriers set up a night ambush position about a kilometre away, on the other side of the previous night's harbour position.

'At about midnight we were awakened by a series of *crump, crump, crump* explosions and quite a bit of light machine-gun fire from a number of AK47 and RPD machine guns. The cavalry radioed from their ambush position to ask if we had a contact. We replied in the negative and asked them the same question; they too replied in the negative.'

The platoon never returned to the old position to check what had happened because they were aware of the potential for booby traps, mines and possible enemy ambush. In any case, they had a big job to do to finish the demolition of the bunker system. But John was convinced that the mortar rounds and automatic weapon fire were from a 3 NVA Regiment group putting in a night attack on their former harbour position after a close reconnaissance the night they heard the clicking.

'Years later, I met the patrol commander of an SAS patrol that had been monitoring the 3 NVA Regiment bunker system,' says John. 'He was amazed that we hadn't encountered NVA strays during our demolition task, as they had continued to move around the bunker system for days after the initial attack and demolition. I just write it all off as a bullet dodged!'

Admirable caution and good military practice had saved John's troops from a pounding. Sometimes, however, especially when you are in a high state of alertness, you can be too aware of potential dangers. And when blokes whose day jobs were to drive trucks, tractors and bulldozers are suddenly expected to operate as infantry, anything can happen.

In Nui Dat, one night in February 1969, five shadowy shapes were seen by infantry night sentries moving rapidly across the front of the Plant Troop position. They were described as dressed in black, moving in quick short bounds and hunched over very close to the ground, clearly taking advantage of the long grass in front of the perimeter wire. The Task Force 'stood to', taking up defensive positions and watching and listening intently for noise or movement outside the perimeter.

Captain Bob Fisher, second-in-command of the Sappers, instructed Lieutenant George Hulse to put together a fighting patrol immediately and move into the area and engage the five enemy. The 12-man fighting patrol was to move out through the Plant Troop machine-gun observation post (OP) nicknamed 'the Taj Mahal', and then to sweep up in front of another engineer machine-gun OP nicknamed 'Kamikaze' because it would be suicide to be in front of it should the engineer crew open fire. Finally, they were to force the enemy into a minefield in front of engineer machine-gun post 'Whisky 2', where they could be captured or eliminated. At least, that was the plan.

'I had to remonstrate with one of my sergeants in the Taj Mahal, who had clearly been drinking more than his ration of two cans per day,' recalls George. 'I threatened to ram the M60 bolts into the machine-gunner's arse if he opened fire while the patrol was out in front.' However, Corporal 'Snow'

Edmunds was way ahead of George. He leaned out of the OP and said, 'Don't worry, boss, I've got 'em both here,' and in the moonlight he showed both M60 bolts in his hand.

The patrol of 12 fired-up plant operators—all quite keen to get into a fight—moved out of the wire and towards the enemy. There was a Land Rover track running at right angles to the perimeter and about 80 metres from the engineer minefield. 'I decided that this line would act as the "line of departure" for an assault on the enemy, with the aim of pushing them into the minefield,' recalls George. 'We had surprise on our side because we had moved very quickly and quietly into our assault position.'

George radioed Captain Fisher that the assault was about to be launched, and Fisher informed him that the tanks of C Squadron, 1st Armoured Regiment were standing by on an adjacent hill with white light and direct fire support if he needed it. 'I was okay with the white light but apprehensive about collateral damage, should the tankies open up with their main armament or machine guns,' says George. 'Anyway, just before I gave the command to attack, I pulled my field glasses out to try to get a fix on the enemy's position. They were still moving under the heads of grass; I remember thinking that their field craft was superb.'

George raised himself higher, trying get his fix, before his driver and radio operator, Sapper Robert 'Shorty' Yates, grabbed his belt and pulled him down, whispering, 'For Christ's sake, George, keep your bloody head down.' They were on first name basis in the bush, although it was always 'Boss' once they were back inside the wire. George took the advice then ordered the patrol to stand and prepare to advance. Before they could take a step, the five enemy emerged from the shadows and onto the Land Rover track—revealing themselves to be five peacocks. On seeing the patrol, the birds wisely turned south and ran for their lives. Not a shot was fired, and George radioed Captain Fisher to report the outcome.

On the patrol's return to the command post, there was much speculation about why peacocks would move together, and at night. It seemed totally against what anybody knew of these solitary male birds, which tend to be active only during the day. The mystery of the Peacock Patrol has never been explained, says George.

More often, however, mistakes made in the field could have serious consequences. Trevor Shelley and his infantry comrades were right at the end of an operation and were anticipating their return to base with considerable relish. The troop had come down from the hills to a scrubby area beside Highway One, and everybody was looking forward to the minimal luxuries of Nui Dat, like sitting on a toilet seat, having a shower and reading their mail from home. As they came out onto the flatland, the platoon commander told the Sappers that a splinter team would sweep and clear an open, well-grassed area so that they could wait to be picked up by both APCs and choppers.

'We had no mine-detectors or anything else, so we just had to concentrate and look for devices or trip-wires, mainly,' says Trevor. 'We thought that we had it sorted after about 30 minutes prowling around, so I gave the lieutenant the all-clear and then squatted down to wait for the rest to move into position.' The infantry emerged from the tall timber and fanned out, on full alert and ready for action.' It was a chilling thrill for the Sappers to see the infantry move towards them like that, from a point of view more often observed by the enemy. They secretly prayed that the grunts remembered they were all on the same side.

'They had come about 70 metres into the clear area when there was a thump and they all went down,' says Trevor. 'There were some wisps of smoke, and when I looked again somebody was rolling around on the ground very close to us. We had missed a trip-wire attached to a hand grenade shoved into a tin. There was no time to think at that stage—just get to the bloke on the ground. Things did not look real good, anyhow, and the amount of blood in his crotch area and upper legs was quite dramatic. Trying to calm him down, I soon realised that his main worry was his family jewels.

'Out with the trusty Sapper's knife, and I said, "I'll check it for you." The grenade must have gone off around crotch level, as there and the upper legs were bleeding. The trousers were shredded anyway, so I cut right across the tops of the trouser legs and bared everything—lo and behold, his important equipment was intact. However, there was one large lump of shrapnel that must have been running out of puff when it got to him. It had carved a very neat groove along the side of his member and came to rest just under the head. There was no real damage and he got medi-vacced soon after, on the first leg of his trip back to Australia. We might have just done him a good turn by missing the grenade.'

Trevor must have been feeling like a good luck charm that day. Later on, a Bell bubble helicopter crashed on the road with the quartermaster of 6 RAR and the pilot in it. 'The QM walked away unhurt, but the pilot was found to have been shot in the head, which is what brought the chopper down,' says Trevor. 'I heard later that, amazingly, he recovered.'

From operating as infantry—like the planties on the Peacock Patrol—to purely performing their splinter team duties, the Sappers had a key role in most of the operations in which they took part. Early in 1970, Sandy MacGregor's brother Chris was involved in an incident that revealed the true nature of the Sapper-infantry relationship. Chris was out with 8 RAR in the Long Hais, in a splinter team supporting a platoon. A couple of weeks before this operation, a very successful ambush position had been put in by Lieutenant Lauder's platoon on what became known as Lauder's Hill; many Vietcong had been killed and wounded.

The platoon Chris was with had been given the task of putting in another ambush on Lauder's Hill. Not far from the objective, Chris warned the platoon commander that the area would be mined—most probably with M16s, the deadly American 'Jumping Jack' mines, which the Vietcong were recycling and using against the Allies. 'How do you know?' was the response. Chris advised that, from experience, the Vietcong would always mine areas where they had suffered casualties. The mines would be along approach paths, in the actual

KNOW YOUR ENEMY

position itself and in the likely areas that a dust-off chopper would come in to evacuate casualties.

The platoon commander spoke to his company commander and was ordered to proceed as planned and occupy the position... except that they should let the Sappers of the splinter team clear the way first. And so the Sappers led the advance, covered and supported by the infantry scouts. As Chris had predicted, he found an M16 booby trap in the ambush position, and the next morning he found two more in the closest landing zone.

Life and death, near-misses and dodged bullets, false alarms and smart soldiering—these were all part of the infantryman's daily life, and were shared by the Sappers of the splinter teams. They weren't merely up there at the sharp end—they were often right in front, clearing the way.

10. LANDMINES DON'T TAKE SIDES

It has been described as Australia's biggest tactical error of the Vietnam War. The Barrier Minefield was one of those 'brilliant' ideas that made sense in theory—but only in theory. The idea was to lay a minefield between the Vietcong's stronghold in the Long Hai mountains and the villages that were their source of food, information and manpower. In early 1967, Brigadier Stuart Graham, the new commander of the Australian Task Force, drew up plans for the Barrier Minefield, more than ten kilometres long and containing 23,000 American Jumping Jack M16 mines. It would turn out to be an own-goal of catastrophic proportions.

Seven Australians died and ten were wounded during the minefield's construction, but that was far from the worst of it. The Vietcong, having worked out a way of removing and redeploying the mines—even after the Sappers had devised an anti-lift device—used the unguarded minefield as a munitions supermarket. M16 mines, most of which were recycled from the Barrier Minefield, eventually killed 77 Australians and injured another 567—more than 20 per cent of the Australian casualties during the entire war.

Sappers made up the biggest proportion of the dead and wounded, even though, ironically, they had been against the idea from the beginning. 'We thought it was bloody madness,' says Sandy MacGregor. 'A minefield is only truly effective if it's covered by small-arms fire from both static and mobile patrols. High-ranking engineers asked who was going to cover such a huge minefield, because we didn't have the resources; the answer was that the ARVN had agreed to do it. We knew right away that that wasn't going to happen, for many reasons. If experience had taught us anything, it was that there were elements of the South Vietnamese forces that couldn't be relied upon, for the simple reason that they were working for the other side. But all of this was ignored.'

Brian Florence, commander of 1 Field Squadron and the recipient of one of only two Military Crosses awarded to engineers (the other went to Sandy), spoke strongly against the plan at Task Force headquarters, however it was decided that the Barrier Minefield should proceed. The Sappers were given the initial task of laying the mines. Ironically, once the mistake was realised, they also had the perilous task of clearing them, a process that would take three years. In the interim, scores of Australians died from their own devices, which had been recycled by the enemy. Bob Coker, who runs the website thecasualtylist.com, says mines accounted for 77 of 365 deaths in the Task Force as a whole, and for 567 of 2515 wounded. Landmines don't take sides.

One would hope there is a special corner of hell reserved for the genius who invented the Jumping Jack. It was a two-stage anti-personnel mine that, when stepped on, would pop up a metre off ground—about waist height for an average man—and then explode sideways, maximising its devastating effect on anyone nearby.

Had the Barrier Minefield been protected, it would have been a formidable obstacle, and there was considerable support for the idea. 'To put it into context with other events at the time, there was a huge minefield across the demilitarised zone between North and South Vietnam, and for all the right reasons it seemed to be working,' recalls Lieutenant Joe Cazey, who was given the job of laying the initial mines. It was decided to build a semi-permanent base in 'the Horseshoe', a small volcanic crater with one open side near Dat Do. 'It was large enough to absorb a company-sized force and to defend it against serious attack,' says Joe. 'A number of substantial bunkers were to be constructed in the style of those around the final positions of the Korean War.'

The Barrier Minefield required a major effort from engineers, so Danny Simpkin of 17 Construction Squadron moved his troop out there. Work also started on digging a decent well so that water didn't have to be trucked or flown in on transports, which might be better employed carrying ammunition and food. As soon as the Horseshoe was secure, a broad swathe was cut through the bush towards the coast, and work parties started to build a catwire fence—two reams of circular barbed wire on top of each other, between steel pickets with an apron of strands of barbed wire from the top and middle fastened into the ground—on either side of where the minefield would be laid.

'I took a big part of 1 Troop on to Horseshoe towards the middle of April 1967 to lay a protective minefield on the saddle of the Horseshoe, facing towards the north, and another to fill the gap across the two shoulders of the feature, facing south,' says Joe. 'These were pretty standard minefields, using the M16 anti-personnel mines.' The first minefield was a useful training exercise before the main task and went in without incident, although the ground was steep and rocky. 'It wasn't like doing it on a level field, as we'd done at the School of Military Engineering.'

The second minefield was to be laid in an area previously used by a number of gun batteries, so it was littered with weapon pits, used defence stores and other signs of occupation. Joe's team decided not to move any of this gear: they didn't have the time, or any need to recover the stores. The day after it was finished, complete with fences and warning signs, Battery A of the 2/35th US Artillery, under Captain Glen Eure, moved in to support operations to the east and south.

'I met the battery commander at the gate and pointed out the area that they had previously occupied, which was now an anti-personnel minefield,' Joe recalls. The US artillerymen moved into their new position, and that day Joe's men started laying mines out on the main minefield. He'd decided to work from the town back towards the Horseshoe, again deploying M16 mines, but this time he decided to put anti-lift devices under them.

It was a simple idea: after carefully placing a hand grenade that had been fitted with a pressure switch and instantaneous fuse under the mine, the Sappers removed its pin; thus, the mine was holding the trigger in place. If the mine was removed, the grenade would explode immediately.

'This not only made progress much slower,' says Joe, 'it was also much more dangerous. It was a very delicate operation to hold the mine and arm the anti-lift hand grenade, before arming the mine itself.' Joe's first task on any mine-laying operation was to make sure the centre of each strip could be found again later if and when the field was to be lifted. 'I used land marks that could be located very accurately—to within 30 centimetres—with pacing and bearings taken several times to ensure there were no mistakes.

'Once this was done and recorded, I left Troop Sergeant George Biddlecombe in charge and went back to the Horseshoe, where other troop members were unpacking, testing and preparing mines. As I was being driven by Sapper 'Blue' Willett out of Dat Do towards the Horseshoe, I saw an explosion near our troop lines. We were only about a kilometre from the position at the time, so we drove quickly over the bumpy road to find a small crowd gathered at the wire of the minefield.'

LANDMINES DON'T TAKE SIDES

Two US gunners had ignored the warnings about the minefield and the signage—presumably because there had been no minefield there previously—and one had climbed through the marked fence surrounding the minefield to retrieve some defence stores they'd left behind. He had detonated a mine, and the resulting explosion had seriously injured both him and his mate at the fence line.

Neil Innes, in an interview for *Holdfast* magazine, takes up the story: 'Myself and two other sappers, Ron Forsyth and Al Hammond, grabbed our weapons and ran towards the source of the explosion. I don't know what we expected to find or do once we reached the "contact", but when we reached the location of the explosion, what we found was two wounded American gunners, one lying on the edge of the minefield, the other lying in the minefield ... writhing about and screaming in what was obviously a great deal of pain.

'It didn't take a genius to understand that someone was going to have to enter the minefield to help the wounded guy as there was no way he was going to be able to leave the minefield by his own efforts.'

Ron Forsyth was married with kids, and Al 'Happy' Hammond was engaged. As Neil was single and unattached, they agreed he should be the one to go in. 'The gunner had gone about 20 metres into the minefield,' recalls Neil. 'I entered the minefield and worked my way towards him. As I didn't have a bayonet with me, I cleared my way forward by running my hands over the ground to feel for the mine prongs. As I found each mine, I marked it.'

A SAPPERS' WAR

Joe Cazey carries on the story: 'By the time we arrived, Sapper Neil Innes had gone into the field and was pinning down the man who'd walked on the mine. Sapper Innes was getting the onlookers—me included—to toss him their green cloth hats so he could mark the active mines. This way we could get a stretcher into the wounded man inside and recover him to safety.'

As Neil continues, 'When the medic, David Buckwalter from A Coy 6 RAR, and I reached the man [Private Pardo] he was lying between two mines—one was near his head, the other near his waist. He was moaning and thrashing his arms about and I had visions of him setting off the mine near his waist so we stuck his hands down his pants and tightened his belt up to restrict his movements. The medic hit him with an injection while I proceeded to mark a clear lane into the minefield to allow further medical aid to come forward and extract Private Pardo.'

The gunner who was on the edge of the minefield, Sergeant Danny Hayes, died from his wounds as he was being placed on the dust-off helicopter. 'One image that sticks in my mind,' Neil writes, 'is how the rest of 1 Troop had arrived back at the Horseshoe while the rescue was taking place, and how they, along with the guys from A Coy 6 RAR and the American Artillery Battery stood outside the minefield like a footy crowd watching us.'

Neil Innes was awarded a Military Medal for his actions, and David Buckwalter was Mentioned in Despatches. For Joe Cazey, once the drama was over, his only concern was the laying of the ten-kilometre minefield. 'As progress was slower than the Task Force Command had expected,' he recalls, 'we got more people in from the troop. Even so, we almost never achieved the desired rate of laying 1000 mines a day.'

The reality was that, every morning, instead of resuming work where they'd finished the night before, a lot of time was lost checking the area for any signs that the mines had been interfered with or had gone off as they were intended to do. 'Each night, we'd withdraw to the Horseshoe and know that there was a risk that the mines might have been tampered with while we were away. Doctrine dictates that a minefield must be covered by observation and fire to make it effective, and this had been a major argument against the whole idea of this minefield from the outset.'

An ARVN Regiment was deployed as promised, but when it arrived it was the battered remnants of a unit badly mauled in a firefight up near the demilitarised zone— and now it comprised fewer than 60 riflemen. 'That meant that we had to have a decent size ambush patrol on the "enemy" side of the minefield,' says Joe, 'to deter Charlie from getting in and doing bad things where we were working. However, what was less evident at the time was that the enemy was on both sides of the minefield, as the great majority of residents of Dat Do were Vietcong sympathisers.

'It became evident that Charlie, local sympathisers or dogs, pigs, deer and oxen were getting into where we were working, and some mines were being set off anyway. So, while the sappers and the junior NCOs got on with their laying drills, it was my job, along with Sergeant Brett Nolan, to retrace the strips' centre lines and re-lay mines with anti-lift devices where the holes were from the detonated mines. It was tense work and there were some close shaves. Brett was a good operator and kept me from doing silly things.

'After a number of days when we got more and more mines laid each day, we had a terrible accident. As Brett and I went into the field—we were nineteen paces along the centre line—Jethro Thompson, who was giving Brett some lip, stepped on the mine he'd just laid. The mine arming party were all around him and the toll was dreadful. When the smoke and dust had settled, there were some six or seven men down, some screaming and others just lying there. I screamed out to call for dust-offs and carefully ran back along the centre line till I was out of the immediate danger area. Assisted by many others, I tried to tend to the injuries as best I could. It was clear Jethro would lose parts of legs and arms but his body and head had been largely protected by the helmet and flak jacket we were all required to wear. Ray Deed died that night of his wounds, and Dennis Brookes died several days later.'

Jethro Thompson would lose his left leg, his right hand and most of the fingers of his left. Despite this, at the time of writing—some 40 years later—he is still going strong, travelling around to catch up with his Sapper mates. Back at the minefield, although the mood on the site was pretty grim, work had to go on. Once the wounded had been evacuated, Joe and his men covered the mess up and went on laying. 'It was tough but there was no point sitting around thinking,' he says.

'Unfortunately, several days later, another man, Glen Bartholomew, who slept with a loaded 0.45 pistol, accidentally shot himself when it discharged sometime in the early hours of the morning. I was standing to at dawn when some of his tent mates came to me and told me that he wouldn't wake up. I went to the tent and found that he was still alive but in a very bad way. He was evacuated but died en route to hospital from loss of blood. The sound of the shot had been muted by individual sandbagged bays each tent was broken up into, so that if we were mortared the casualties would be minimised. The sentry had reported a single shot at about 0230 hours, but he couldn't tell from which direction the shot had been fired and no action was taken.

'Shortly after this, 1 Troop was relieved by 2 Troop, which was to continue laying and finish the stretch to the Horseshoe. They had two fatalities within the first day after they took over from us. Sadly, even more were to die as a result of the Barrier Minefield blunderplan.' Eventually, Joe would achieve his target of 1000 mines a day, but only after infantry were brought in to dig the holes where the mines would be placed.

At first, the anti-lift device was effective in preventing the mines from being stolen—and the Vietcong couldn't work out why mines were blowing up when they had worked out how to replace the safety pin in them before lifting them. But then a Vietcong soldier was carefully removing a mine when he heard a click underneath it. He—or as one story has it, she—had triggered the anti-lift grenade, but it was a dud and hadn't gone off.

A SAPPERS' WAR

For the first time, the Vietcong had evidence of why their mine-stealers were being blown up, and they were able to neutralise both the mines and the anti-lift devices and reuse them both. As this knowledge spread, the result was an increase in the recycling of the mines to be used against Allied troops. When in 1968 the ATF commanders realised their mistake—not only was the Barrier Minefield ineffective, it was a virtual ordnance giveaway for the Vietcong—Australian engineers were instructed to start clearing it.

Regardless of which side of the fence you sat on regarding war in general and the Vietnam War in particular, there were few more tragic illustrations of the waste of human life than the Barrier Minefield saga. Australian soldiers died laying the mines and died clearing them, and others were victims of an ingenious foe who learned to use their own weapons against them. This was pointedly brought home in an incident that became a song, which, a decade later, forced Australia to confront the uncomfortable truth about the young men, rejected and sometimes reviled, whose only crime was to answer their country's call to arms.

11. THE SONG THAT SHOOK A NATION

Then someone yelled out 'Contact', and the bloke behind me swore.
We hooked in there for hours, then a God almighty roar;
Frankie kicked a mine the day that mankind kicked the moon:—
God help me, he was going home in June.

I can still see Frankie, drinking tinnies in the Grand Hotel
On a thirty-six hour rec. leave in Vung Tau.
And I can still hear Frankie lying screaming in the jungle.
'Till the morphine came and killed the bloody row.

And the Anzac legends didn't mention mud and blood and tears,
And stories that my father told me never seemed quite real
I caught some pieces in my back that I didn't even feel...
God help me, I was only nineteen.

From 'I Was Only Nineteen' by John Schumann

Dave Sturmer was a little older than 19—he had just turned 21 a month before—but he remembers it all vividly. How could he not? It was the day man landed on the moon (although it was still only the 20th in the USA), the day his splinter team partner won a Military Medal, and the day he was in the thick of the action when a legend was forged that would be heard around the world. But considering how close he came to losing his life—he was injured but still had to deal with the devastating effects of the mine blast described in 'I Was Only Nineteen'— it's ironic that, initially, the army pretty much lost Dave.

Dave had no idea he was even going to be part of a splinter team when he'd arrived in Vietnam three months earlier. The army had lost his papers and didn't know what to do with him. 'I started my service in 1968 and got to Vietnam in April 1969,' Dave says, 'and they put me in 55 Workshops, which was no place for a field engineer. They sat me down and said, "Why are you here?" A guy called Sergeant Christie, who I got on with very well, said, "Dave, just sweep out the bloody warehouse because we don't know what to do with you." But he got to the bottom of it and, I must say, I wasn't too happy about going forward. I'd been in Vung Tau for about two weeks and I thought, "Look, I'm surfing, I'm having a good time! I've got brothels over here, I've got good bars—why can't I stay here?"

'That said, there was a lot of boredom, and there were a lot of fights—you got pissed and it was, "Here we go." But the next morning it was all forgotten about. At one stage I think we all thought we were bulletproof, but you soon find out you're not! There are guys who know they're not coming back. A guy asked me to paint a table in a bar with his name on it because he reckoned he wasn't coming back—and he didn't.'

Three months later, Dave nearly didn't come back either. It was 21 July 1969, the day man first landed on the moon. 'I was on Operation Mundingburra when that erupted,' Dave says. 'I was only the number two, learning my trade.' His number one in the two-man splinter team attached to 3 Platoon of 6 RAR was Corporal Phil Baxter, and their platoon commander was Lieutenant Peter Hines. The operation was near the Vietcong stronghold of the Long Hai mountains, so they were on high alert for mines. Dave and Phil's job was to find the mines and either neutralise them or find a way through them. The platoon had just been told that the Long Hais were about to be bombed by American B-52s, and a warning sign for a Vietcong minefield had been spotted.

'We knew we were in a bad area, but we didn't quite know where the mines were,' Dave remembers. 'We were being very careful and moving very slowly, but we didn't see the other signs leading into the track, so we went straight through the middle—otherwise, we would have known there were mines somewhere in that space between the two signs.' It was only by sheer good fortune that the platoon hadn't struck a mine already. That said, it was bad luck that led to Lieutenant Hines triggering the first explosion after he called for a smoko while they waited for the bombing raid to end.

'We were clearing up the tracks and the Louie came up to us and told us the Yanks were on the moon—we didn't even know they were up there,' says Dave. 'Then he stepped over our kit and onto the mine—a Jumping Jack—and that cut him in half. Suddenly we were down to 17 in the platoon in one fell swoop. We had put the mine-detector together right on top of the mine but never trod on it once,' explains Dave. 'You set up your mine-detector next to your foot because army boots have steel plates in them that gave you a good reading to tell you if the detector was working or not. Trouble was, we were also getting a reading from the mine, but we had no idea it was there. For some reason we didn't tread on it.'

Even though they were both injured, Dave and Phil had a job to do. 'My first thought on hearing and feeling the massive explosion was that the B-52s had dropped a bomb short of target—but this wasn't the case,' Phil Baxter told *Holdfast* magazine. 'After the initial shock and the realisation that we were in a mine incident, we yelled out to everyone to remain still. Not to move. Dave and I attended to each other's wounds, then I sent Dave prodding with his bayonet to clear a safe lane towards Lieutenant Hines, while I prodded towards the radio to call in assistance,' Phil recalled.

Dave continues the story. 'People like Phil, myself and an infantry corporal called John Needs ended up taking command and really held things together. Phil was quick to react, and we both started clearing from where we had started—back through the devastation and the path that we knew was safe. We were moving but everyone else was told to stay very still until help could come to them. A piece of shrapnel had pierced a smoke grenade and was burning into the back of our first bloke, and the smoke grenade was flanked by two fragment grenades, so he was freaking out. Once we'd removed the smoke grenade he calmed down, and we moved on to a sight that I could not believe possible. The lieutenant had been really terribly wounded by the mine but was still alive and in control. This brave man not only calmed the wounded close by him, he passed on command to a young corporal. I just couldn't get over how, looking at half a person, he could still take control. He must have been in a lot of shock and not feeling anything.' Lieutenant Hines would die soon after, in Dave Sturmer's arms.

One of the problems the Sappers had was that, being in the middle of a minefield, they not only had to clear paths to the wounded, they also had to tell men who were literally writhing in agony to lie still in case they set off another mine. 'I have to say, as much it's a sad thing, the bravery of each and every one of those guys was just sensational,' says Dave. 'There are so many things I learnt much later about how these guys trained, talking to them over various bits and pieces, the things like battlefield clearance that they had worked on long before they went over. They were such a tight-knit platoon—I guess that held them together.'

The next task was to clear an area so that choppers could both fly in a CET to deal with the rest of the mines and take away the injured men—including the Sappers of the splinter team. Among the men helping to clear the pick-up zone—the choppers never actually landed but just hovered and used winches—was infantryman Mick Storen. Mick also helped load the wounded until one of the medics noticed that Mick's back had been shredded by shrapnel. Mick hadn't even realised, but he too was ordered onto the chopper. But this disastrous day wasn't over yet. There was more mayhem when the regimental medical officer (RMO), Captain Trevor Anderson, and the 6 RAR Battalion's commanding officer, Lieutenant Colonel David Butler, arrived on the scene.

'Trevor Anderson—the doctor—saw the minefield sign and went off the safe lane to take it off the tree,' says Dave. 'That's when he trod on the second mine. It didn't jump out off the ground, it just went off under his feet and sheared his front off.'

This second blast killed Corporal Needs—the infantryman who had done so much to help Phil and Dave—and injured Lieutenant Colonel Butler. Captain Anderson was seriously injured in the stomach and legs; he would lose his sight in both eyes from the incident.

'The last time I saw Trevor he was in a plaster cast—he looked like a mummy,' says Dave. 'That was holding him together till they got him to Butterworth and could do something with him. He only died a couple of years ago but I think he would have lived with the fact that what he did cost the lives of other people. A CET had come in, and we couldn't move, basically, until they had cleared the area.'

After rehabilitation, Trevor Anderson trained in psychiatry at Royal Park Hospital, Melbourne, and the Parkville Psychiatric Unit, studying with the help of his wife, Janice. Despite his loss of sight, he had a long and distinguished career, developing community psychiatry, training young psychiatrists and inspiring medical students. He succumbed to pancreatic cancer in 2004.

Phil Baxter was awarded the Military Medal for his role in the incident. 'With so much going on, you didn't have time to think of anything other than getting out of there without any further casualties,' he says. 'Afterwards, I didn't feel I'd done anything special—I was just using my skills and training to help my mates. I'm naturally proud of the award, but I'm very aware that it was a team effort... I received great help from my number two, Dave Sturmer, and from Corporal John Needs, who was unfortunately killed after I'd been evacuated.'

There are many, Sandy MacGregor foremost among them, who think that but for the patently unfair system of rationing military honours, Dave Sturmer and John Needs would both have won Military Medals too. They had certainly earned them. Dave Sturmer is also full of praise for the guys who stayed out there after the injured were medi-vacced out. 'I thought the blokes that survived and stayed out there and weren't allowed to come back in were amazingly brave,' he says, including the Sappers of the CET who stayed with them to bolster the numbers of the depleted infantry platoon. 'There was a lot of bravery by a lot of people. There were a lot of things I didn't agree with, after the event, but there were a lot of things that you could say, "That was amazing!"'

As for the song, 12 years later, in 1981, songwriter John Schumann met Denny, his future wife, whose brother was Mick Storen. John told Mick he wanted to write a song about Vietnam veterans and asked if he'd be prepared to help with some of the detail, although Denny had warned John he didn't like to talk about his time in Vietnam.

'I was quite surprised when he agreed, but there were two conditions,' says John. 'The first was that I didn't denigrate his mates, and the second was that I played any song I wrote to him first. If he didn't approve, the song was not to see the light of day. I agreed.'

One night Mick brought a carton of beer and a small cardboard box containing his Vietnam memorabilia to John's place. Denny set up her cassette recorder and they filled around nine 60-minute cassettes. John listened to the tapes for hours on end over the next few months. One morning, he took a cup of coffee, his guitar, a pad and a biro into the tiny backyard of his house in Carlton.

'Sometimes songs take months to write,' says John. 'Sometimes they just tumble out. I reckon I wrote "I Was Only Nineteen" in 15 minutes. It was like it'd already been written. As proud as I am of "Nineteen", that morning I felt as if I was little more than a conduit.' When John played the song to Mick at a family barbecue, it got his all-important thumbs-up. 'I Was Only Nineteen' went on to become an anthem for those demanding better treatment of Vietnam veterans.

But what about Frankie, the bloke who 'kicked a mine' in the song? There was a Frankie in the troop—Frank Hunt, the forward scout—but it was the unfortunate, heroic Lieutenant Hines who had accidentally triggered the initial blast. However, in deference to his family, it was decided that the name would be changed.

'From what I remember, Frankie was okay with that,' says John, 'And I know he was fantastic in promoting the song and helping to draw attention to the plight of Vietnam vets. Did he ever think it really was him that kicked the mine? It's hard to say. Things get confused over time and stories change in the retelling. A lot of people thought the whole song was about Frankie but it wasn't... it was written from Mick's point of view, including that line "I caught some pieces in my back that I didn't even feel". But Frankie was there and he was badly hurt in the incident, that's for sure.'

Back in Vietnam, Dave Sturmer would be injured again, this time in 1969 when he was part of a mini-team attached to APCs. But in between times, he went on operations that were so traumatic—even more so than the Mundingburra incident—that his memory has locked them away in a place they can't be accessed. 'The parts missing are generally from August to November,' says Dave. 'I know I was in the bush with the Yanks, and that was just another planet, working with American engineers. The drugs were rampant, the racial prejudice was rampant. In fact, that was where I learnt about racial discrimination.'

Working with the Americans may, indeed, have been 'a different planet', but for Dave Sturmer, the day man landed on the moon was the day he truly came down to earth.

12. WITH TANKS . . .

Although Sappers were virtually joined at the hip with infantry, under the mini-team system they also had a mutually beneficial relationship with armoured units. The plus for the tankies and APC crews was that they had the services of Sappers, complete with a mine-detector, wherever and whenever they were required. The Sappers were, in the main, sitting on top of the armoured vehicles, watching the road ahead for tell-tale signs of trouble. Anything sighted meant that the driver had to react quickly. Then the mini-team would go into action and clear the suspicious package, often a mine, and sweep the road ahead of them for potential danger.

The upside for the Sappers was that, instead of trudging around the bush with the grunts, they got to ride around on tanks and APCs. 'I found it almost unbelievable that Sappers would be on the outside of APCs with only their flak jackets for protection,' says Sandy, 'but they reckoned that was safer than being inside.' However, it was not without its dangers, as was discovered by Sapper Dave Sturmer, who was injured three times—twice in one day. He didn't know whether he was unlucky to get hit or lucky to survive.

The first time Dave was hurt was in the incident that inspired 'I Was Only Nineteen'. The second and third times were on 8 December 1969 during Operation Marsden, a 6 RAR joint mission west of the Nui May Tao mountains. APCs were sent to assist with the removal of a temporary bridge (and the bridge-laying tank), and Dave, recently elevated to the rank of corporal, was part of a mini-team. Their job was to sit on top of the APC near the driver and look out for possible mines.

'We were mine-clearing, so you were always up front, and you sat with a couple of flak jackets under you so you didn't lose your balls,' Dave laughs. 'You're in front of the convoy, in front of a column going anywhere. You had a job to do so you went and did it. I got hit twice on the same day. They let us in but they didn't let us out.

'I was on top of an APC when we hit the first of three mines—a very large one that picked the carrier up. I stayed with the carrier as it lifted but then it sat down on an even bigger mine. That was when I got blown off the carrier, and that's probably what saved my life. My number two, Sapper John Greene, went straight up in the air, and as the second mine went off he came down directly above it. He was killed, of course. The driver was blown out of the vehicle—as he came out, the hatch actually sheared his hands off. He was in an extremely bad way.

'I was blown a good 50 metres up the track and lost all my clothing... everything had blown off me. The poor old group commander, Graham Locke, was like a singed banana from the waist up, he wasn't real flash at all but he lived for a few years—he died only recently, in fact. I went up and had a chat with the boys, and they said there was nothing I could have done. Maybe about an hour or so after that, a medivac chopper came in—it came down a bit hard so the rotor blade at the back set off a mine that blew the tail off the chopper and I got hurt again. The last thing I remember about that is that the chopper crew broke out their life raft and offered us water and bickies from Oz—they were having quite a good time.

'By that stage, I had to make my own way back on the APCs because we only had a limited number of people that could get up in the chopper. The Vietcong were still out there—the second carrier had seen movement about 100 metres into the scrub. They were hooking in on them while I was on the ground still trying to find out whether I still had any pieces on me. Our guys were firing into the bush. It was a strange place to be, and I kept praying that they didn't lower the 50-calibre heavy machine gun that was mounted on the APC.'

One heroic member of a mini-team was not so lucky. When 4 Troop, B Squadron, 1st Armoured Regiment set out on Operation Chieftain to patrol and ambush in the Duc Thanh district on 3 September 1969, they took along a mini-team of two Sappers from 1 Troop. Sapper Tony Lisle was the number

WITH TANKS...

two in the team. Intelligence sources had indicated that the Vietcong's D440 Local Force Battalion planned to attack and harass on Route 2 during the coming month. So ambushes were to be set up in areas near the road.

In his book on the history of the 1st Armoured Regiment in Vietnam, Bruce Cameron MC describes what happened just three days into the operation: 'Sapper Tony Lisle typified the mini-team members with whom the tanks were privileged to work—someone who was enthusiastic, responsible and, above all, proud of his work. During Operation Chieftain he was riding on the track guard of the lead tank when suddenly he screamed out and hit the driver on the head (the signal to stop immediately!). The operator, Trooper "Winga" Williams, distinctly remembers the urgency in Lisle's voice. His warning had the desired effect. The driver, Trooper "Moe" Burgess, must have pulled both sticks and had both feet on the brake, because he nearly stood the tank on its nose.'

Tony Lisle had noticed two things. The first was a short length of wire that turned out to be running from a pressure switch to about 25 kilograms of Chicom explosive. The mine was set in old tank tracks that had been cleverly reshaped to match the rest of the track marks, in order to disguise the fact that explosives had been buried underneath. Luckily, a very small piece of wire had been exposed by rain. Tony had also noticed that the grass on the side of the road had been arranged in an 'X'. This turned out to be a Vietcong mine marker, showing the location of the pressure switch. 'The alertness and quick reaction of the mini-team member brought the tank to a halt within centimetres of the switch,' wrote Cameron. 'Burgess had to reverse to enable the explosive to be dug up.'

Tanks had travelled this road in the opposite direction the day before; because it was the wet season, they had left distinct track marks. Later, Tony Lisle explained to the crew how the enemy had planted the mine and switch and then refashioned the tank tracks by hand so it appeared that a tank had already driven over the spot where the mine was buried.

'Many people would have relaxed somewhat, having made such a discovery, not so the mini-team,' Cameron continues. 'Twenty metres or so further up the road was another (as yet undiscovered) mine. When the tank stopped for the first mine, Sapper Lisle jumped off into the centre of the road and waved the tank to back up a few metres. When it did so, he then bent down and with a couple of flicks of his bayonet, exposed the pressure plate. The mini-team then disarmed the mine and dug the explosive out of the ground. While we were "ohhing and ahhing", Lisle wandered off further up the road and found the second mine. This one contained 23 kilograms of explosive and was again laid in old tank tracks.'

Four Troop and the crew of 24 Alpha—Second Lieutenant Chris Sweeney, Lance Corporal Bob Ferrari, 'Winga' Williams, 'Moe' Burgess and gunner Chris 'Java' Jones—were both impressed and grateful. 'I could have kissed Tony (but I didn't),' says Chris Sweeney. 'He had definitely saved 4 Troop from taking casualties that day. I told Tony that next time we were in Nui Dat I would buy him all the beer he could drink. Unfortunately, this never happened.'

WITH TANKS...

Just six weeks after this incident, Tony was in a mini-team attached to B Squadron 3 Cavalry, in a joint Australian/ARVN operation to the east of the area known as the 'Light Greens', when Corporal Fred Venema's APC was ambushed. With rocket-propelled grenades (RPGs) hitting the front of the vehicle and the turret, the driver, Trooper Bert Casey, was killed; Fred and Tony were wounded. Tony died from his wounds in 1 Australian Field Hospital, Vung Tau, 20 days later, leaving a two-year-old daughter, Julie, who would barely recall him, and his wife, Helen, who was pregnant with their second daughter, Kathryn, who would never know him. But those tankies of 24 Alpha would never forget him.

If APCs were vulnerable, you'd think you'd be reasonably safe on or in a tank. 'Up to a point,' would be the response of Clive Pearsall, a Sapper with 1 Troop from 1967 to 1968. 'I was riding in a convoy of three Centurion tanks,' says Clive, 'sitting on top of the second tank, along with my number two, Ken Wheatley, who was out on his first operation. As we hurtled along at about 40 kilometres per hour, the tank in front hit a massive mine. In slow motion, I saw the lead tank lurch up off the ground at about 45 degrees. Then there was the shuddering roar of the blast, and both tanks came to a halt as dust, stones and rocks went flying 50 feet into the air. Incredibly, nobody was badly injured—and afterwards Ken and I joked how we were so black from blast debris that we looked like coalminers.'

Michael Bidey, who Clive met at a Vietnam veterans' lunch decades later, according to an article in *Holdfast* magazine, was the gunner of the Centurion tank that hit the mine. 'The 3rd of August, 1968, who could forget it,' he writes. 'The crew of 31A was Sgt Nev Callis, Kevin Hunter (driver), Barry Munari (operator) and me as the gunner. I was filling in for the regular gunner who was on R&R. It happened north of Nui Dat not far from FSB Avenger in what I seem to remember was an area known as Courtney Rubber. I can to this day, close my eyes and see myself being thrown around the inside of the tank in slow motion—first whacking my head on the turret roof and then crashing down onto my backside. It was like a dust storm inside the turret.

'We were very lucky—I had bruises everywhere, a crushing headache and a sore back. It was either this blast or the next mine in September where I cracked my pelvic bone. My first reaction was that we had been hit by a rocket—not that I knew what being hit by a rocket was like. But it seemed to me that we had been hit on our right side and I immediately swung the turret to the right.

'The crew commander (Callis), yelled at me to traverse back as we were aiming at one of our tanks. This was the largest mine encountered by the Squadron since it arrived in country earlier that year. It left a crater about 12 feet wide and over 4 feet deep. I had heard that it was about 20 kilograms of Chicom TNT and it lifted the tank clear off the ground. There was extensive damage to the tank—parts of the track were blown all over the place—the final drive, right rear suspension and track wheels were blown off.'

Fortunately, they had an armoured recovery vehicle (ARV)

WITH TANKS...

with them, and incredibly, the mechanics managed to get the tank up and going again. It had been blown to an angle of 45 degrees, defying the accepted wisdom that no one inside could survive a blast strong enough to lift a 52-tonne tank.

Most outings with the tankies had some sort of high excitement along the way, interspersed with bone-jolting journeys through rough country, and periods of sitting and waiting for the fun and games to start all over again. But whenever Sappers were deployed—and whatever their role was—the challenges were usually deadly serious. In February 1969, two battalions from the Thu Duc Vietcong Regiment were thought to have entered the Hat Dich area, north-west of the Task Force base. Operation Goodwood was launched, unusually at that time involving both Australian and American forces. Armour, Infantry and the Sappers were all involved, including Victor and Whiskey companies, which were New Zealand infantry companies forming part of 4 RAR.

Australian Sappers Bob Laird and Kerry McCormack formed a splinter team with Whiskey Company, while Sappers Rob Woolley and Geoff Anderson comprised a mini-team travelling on the tank-recovery vehicle. A US Liaison Officer was involved, who had an interpreter from the Hoi Chuan Program for prisoners-of-war who had turned over to the Allies. While B Company began clearing the area, forward scout Barry Butler realised he was in an enemy fire lane but before he could take evasive action a huge explosion occurred. He was blown sideways and soldiers behind him were killed and wounded, including the Company HQ Group. The tanks from 1 Troop

(Armour) came forward but the Recovery Vehicle stayed out of the battle. A US helicopter gunship was overhead, firing on enemy lines but with all the confusion of battle it also mistakenly fired on Allied forces. To add to the mayhem, the grass around the area was very dry and spot fires were springing up all around the action.

The Centurion tank 21-Alpha was firing canister (fragmentation) rounds to clear bunkers but they ran out of canister and started firing both their 30-calibre and 50-calibre machine guns, which then jammed at the same time. Taking advantage of the tank crew's inability to defend themselves, some Vietcong and the NVA regulars jumped up with an RPG-7 and the tank took several direct hits. The tank commander and his crew were all injured and were told to abandon the tank. A radio call to help rescue the wounded was heard by Rob Woolley and Geoff Anderson, who got off the recovery vehicle, made a stretcher from Geoff's hootchie and, unarmed, went forward to assist. An Infantry Sergeant also put his own life at risk to pull the wounded crew out while another tank fired to keep the enemy's heads down. Rob and Geoff helped Medic Sam Brown deal with the wounded. Soon after the doctor arrived by helicopter and Rob Woolley guided him to the scene. Brown was awarded the MM for his part in the action.

The tanks and Infantry withdrew from the area, with the damaged tank being dragged out, and Phantom bomber planes came in with napalm. Canberra bombers continued during the night with bombs and napalm and by next morning the enemy had gone. However, they must have taken a few casualties because there were quite a few bodies in the area, meaning they hadn't been able to remove all their dead, as they usually did.

Sapper Bob Hamblyn remembers every detail of one patrol with armour in June 1971: 'Sapper John "Crossy" Cross came to my hooch and said, "We are going on this op with C Squadron tanks—we will sleep in their lines tonight as it's 'boots and saddles' at 0200 hours." On arriving at C Squadron we headed directly to the boozer for a couple of cans of beer, a game or two of darts, a surprise meet-up with a schoolmate, Andy Hayman, the unit cook, and then the fartsack' (meaning, bed).

At 2 am next morning, they and their gear were loaded onto a Centurion tank. But it hadn't gone 100 metres before it got bogged, so they mounted another Centurion and were away, with light beacons showing the way up Route 2 towards Blackhorse, an American FSB. 'Looking back at the lights in Nui Dat, I couldn't help but think of Luna Park in Sydney, for some reason,' remembers Bob. 'We passed the rubbish dump of Blackhorse shortly after daybreak, and then it was on into Long Khan province. We travelled most of the day, through the occasional village, where kids would come to run in the tracks made by the armour. Once through the villages, derelict APCs came into sight and were a visual reminder of where we were—and our task ahead.'

At some point, Sappers Cross and Hamblyn separated; Crossy was on the ARV but Bob was still on the Centurion tank. It was only when the Centurion got involved in fairly heavy scrub-bashing through bush sporting thorns 'not unlike three-inch jolt-head nails' that Bob joined Crossy on the ARV, leaving the mine-detector on the Centurion. They continued through the scrub until shortly before sundown, when the platoon set up a night harbour position, with the Sappers placing Claymore mines and sound-detection devices around the perimeter.

The next morning, the mini-team members were retrieving their Claymores and early-warning listening devices when the platoon was ordered into action. During the night they'd heard radio chatter that scouts from 3 RAR had sighted a number of enemy moving through and around the Vietcong's bunker systems. 'I had come off the last picket duty and stood down with the rest,' says Bob, who was listening to the driver of the ARV, telling a story about a tankie who, instead of going back under his tank to replace the floor drain plug during maintenance, had decided to screw the floor plug in from inside. It would have been all right, the ARV driver explained, except the tank hit a land mine and the blast sheared the fine thread of the drainage plug turning it into a lethal projectile, shredding its way through the driver's seat and killing him instantly.

'Then he pointed to my just-made brew and said, "No time for that now—we're going in." I tipped out a full metal cup of tea without having taken a sip.'

The platoon had spent the night on the edge of fairly spindly vegetation that was easily pushed aside by the armour, and the further they went, the sparser it got, until they found themselves in a small clearing. And that was when the fun started. Suddenly they were under attack from RPG assaults

WITH TANKS...

and small-arms fire from enemy bunkers. The first explosion slammed the ARV's rear hatch door shut.

'Within seconds the hatch door opened again,' recalls Bob. 'Pushed toward me from the inside was a 7.62 automatic rifle and three 30-round magazines with the command, "Use this... Watch it—it's been tricked up to fully automatic." The hatch door slammed shut again and stayed that way for the most part of the day. The armour halted, tracked 90 degrees to the right, then advanced on the enemy in their bunkers, all the while firing both solid shot and canister from the 75-mm cannon, as well as 50- and 30-calibre machine guns.

'I lay prone at the forward quarter space opening of the ARV, from where I observed Second Lieutenant Bruce Cameron, commanding officer of 5 Troop, C Squadron, Royal Australian Armoured Corps, his head adorned with beret and headphones, directing his troop and bringing fire to bear down on the enemy in their bunkers. From the ARV, I provided aggressive supportive fire down Bruce Cameron's left flank.'

Bob's automatic rifle jammed after about 40 rounds on automatic—probably a gas stoppage from overheating. Bob followed 'immediate action' procedure, removing the magazine and clipping it onto his 7.62 self-loading rifle (SLR), then continuing to fire single shots into the scrub down Bruce Cameron's left flank.

'At this time I could just discern, over the battle noise, Crossy hollering at me, "What are you shooting at?" I saw no point in trying to respond as I could barely make out what he was saying. Crossy had adopted a half-lying, half-seated position, with his shoulders against the ARV's sidewall, and he was pulling his cloth bush hat down over his face—in the hope that all this was going to go away, I expect. It did not. I just pointed towards the bunkers and said no more.

'Using my SLR, I fired a further 80 rounds of my own ammunition. When I realised that Crossy was carrying an M16 (Colt Armalite) 5.56-calibre rifle—which was incompatible with my 7.62 SLR—I stopped firing. I had only four full 20-round magazines left out of the eight I had started with—just eighty rounds—and I didn't know when they might be needed, since the battle was far from over.'

Bob remembers the next phase of the battle as an ordeal of extreme discomfort and no little fear. He and Crossy were now lying completely flat on their bellies on the ARV's rear, baked by the heat from its engine. When the heat became unbearable, they'd simply roll on to their backs until that was too much, then roll back and forth until it was all over. If they needed to take a piss, they did so where they lay, the engine's heat evaporating the urine almost immediately. But that was the least of their worries.

'Many years later, speaking with Crossy, it dawned on us that we both feared the likelihood of a satchel charge or grenades being hurled onto the back of the ARV,' Bob says. 'Needless to say, we spent more time on our backs than our stomachs preparing for such an event.'

With the seemingly incessant battle sounds now deafening, the ARV began slewing on the bunkers' overhead protection. 'I could see Cameron's Centurion doing likewise and, on firing

WITH TANKS...

canister, almost disappearing in the dust, foliage and smoke,' says Bob. 'Enemy fire from the bunkers seemed to come from all over the place, but they had trouble bringing fire to bear on the tanks and ARV from their bunker slits. This would explain the rounds flicking audibly through the lower canopy of the bush above our heads. The ARV received fire like this for a number of hours, then we left the section and moved out to a small resupply base for fuel and ammunition, where again I briefly caught up with Andy Hayman. Once resupplied, the ARV rejoined the section and the battle came to an end soon after.'

It soon became apparent that Bob had suffered an unexpected loss: an RPG had slammed into the jerry can basket frame at the rear of the Centurion, blasting the mine-detector from where he had packed it earlier. He had been travelling on this tank the previous day and had neglected to transfer the mine-detector with him over to the ARV. But there is a universal response to such crises, as Bob relates. 'Crossy said, "What do you reckon? Brew up?" I hadn't had a brew all day so I didn't have to be asked twice.'

Bob and Crossy selected as shady a spot as they could find alongside a Centurion. They dropped their packs and had started to rummage through their rations when they were hit by the sudden stench of death. The tank was parked right on top of the remains of a long-dead enemy soldier. Ray 'Shorty' Fulton, attached to Delta Company, 3 RAR, picked up the remains by the trousers and removed them for burial.

That night, an 'ordnance mangled tree', as Bob describes it, fell down, pinning a night picket by the legs to the top of an armoured vehicle. Medics treated him throughout the night until he could be medi-vacced out at daybreak. Among other general Sapper tasks over the next few days, Bob and Crossy lent some C4 explosives and det-cord to Ziggy Gniot and Keith Burley, the splinter team attached to the infantry Bravo Company, and helped them bring down an enormous teak tree that was impeding the flight of choppers in and out of the newly constructed landing zone.

But the adrenaline-pumping excitement of a pitched battle was soon behind them and it was time to strike camp and return to Nui Dat. Somewhere north of the De Courtney rubber plantation on Route 2, the ARV took a Centurion with a seized transmission into tow. 'The tank's gears were placed in neutral and it was hooked up to an A-frame tow hitched to the ARV and we set off,' says Bob. 'Coming around a sweeping bend, the ARV left the tar seal and tracked onto the soft edge of the road, breaking the A-frame, which resulted in a teeth-grinding and bone-jolting jack-knife at approximately 40 miles per hour, with the Centurion's 75-mm barrel swishing over our heads. Both armoured vehicles weighed in at around fifty tonnes each. It was not until the early hours of the morning that 1 Field Transport picked us up and returned us to Nui Dat, by which time I was eagerly looking forward to a hot shower and lying prone on a mattress... any mattress.'

Safely back with 2 Troop behind the wire at Nui Dat, Bob was ordered by John Cross and an increasingly annoyed officer in command, Captain John Tick, to fill out the appropriate

WITH TANKS ...

paperwork—an L&D (loss and damage) report—to account for the missing mine-detector. Bob put it off for as long as he could. 'I recall irritably thinking, "The bloody NVA were responsible for the missing mine-detector—let them fill out the L&D report",' he says. The same day that Bob reluctantly filled out the report, a 3 Cavalry Officer returned a severely damaged mine-detector case that had been driven over and crushed by a track vehicle. Bob never did get around to changing the L&D report from 'lost' to 'damaged'.

13. BLACK SATURDAY

By the beginning of 1970, the Vietcong's cunning recycling of American-made mines for use against Australian troops was in full swing and it would have an impact that went well beyond the carnage and devastation of the battlefield. At that time, public opinion in Australia was still, on balance, in favour of the Vietnam War, but it was waning a little. However, then came Black Saturday, one of the darkest days in Sapper history, and an incident that would prompt a sea change in the political debate over Australia's involvement in Vietnam.

On 28 February 1970, a platoon that was part of Operation Hammersley—an attempt to take on the Vietcong on their own patch—walked into an enemy minefield containing strategically placed recycled M16 landmines, doubtless stolen from the Barrier Minefield. Sappers from the splinter team attached to the platoon were trying to defuse a booby trap when they triggered a landmine that had been placed nearby for just such an opportunity. By the time the day was over, nine Australians had died out of a total of 35 casualties. Among the dead and injured were seven Sappers from 3 Troop.

Black Saturday has a special significance for Sandy MacGregor because his brother Chris was right in the thick of it. It was mines that did the damage, and when one member of the original splinter team was killed and the other injured, Chris MacGregor was one of the engineers flown in to find a way through the waiting mines and booby traps to the wounded men without killing themselves. Remarkably, he has no memory of the incident; like many Vietnam veterans, his mind has drawn a veil across recollections that are best kept hidden from their bearer.

It had all started on 15 February, when 9 Platoon, C Company, 8 RAR left FSB Isa to conduct an ambush at a point where members of the Vietcong D445 Battalion were believed to cross a firebreak as they moved out of their stronghold in the Long Hai mountains—often to plunder local villages for food, ammunition and even recruits, as well as to conduct operations against the Allies. It was suspected that Jumping Jacks from the Barrier Minefield had been replanted in this area in great numbers to deter Australian Task Force incursions. So far it had worked, but Australian commanders were unhappy at having a seemingly impregnable Vietcong stronghold right in their midst, and the decision was made to test the enemy's mettle in their own backyard, which until then had been effectively a no-go area.

A splinter team from 3 Troop was attached to 9 Platoon, and their job was to use mine-detectors to establish a safe path to the ambush point from their drop-off zone. However, because the area was a known enemy stronghold, it had been so heavily and consistently bombed and shelled over the years that there were hundreds of false positives—the detectors found shrapnel and bullets in almost every sweep. It took hours to cover just 500 metres because no one would take a step until the Sappers had cleared the way.

The ambush point was a hill overlooking a clearing in a firebreak trail. Defences were established, including Claymore mines, in case the intended victims of the ambush were able to counter-attack. The Sappers who had been out setting up the Claymore arrays had sensibly disconnected the firing clickers while they worked. Unfortunately, they were on their way back when a large group of enemy fighters was sighted.

It might have been a disaster: the Australians were vastly outnumbered and had been forced to spring their trap before they were ready. It took several minutes before the Sappers were out of harm's way and the Claymores were reconnected and fired; after a five-hour firefight, tanks brought reinforcements and drove the enemy off. In the end, the Vietcong suffered 35 deaths, while only nine Australians were injured, none seriously.

The success of this operation emboldened the Australian commanders, who decided it was time to take back the Long Hais—or at least make a dent in the Vietcong's apparent free rein in the area. The new operation was codenamed

The Australian landing ship medium (LSM) Clive Steele unloading in Vung Tau. Sappers were in charge of ships too. The LSMs also delivered the tanks upriver to Baria. Photo by Peter Coppleman/AWM neg no. P05923.008

A typical one-stop shop for soldiers on leave in Vung Tau—with a barber shop and photo processing downstairs, and a massage parlour and steam bath upstairs. Photo by Mike Dwyer/AWM neg no. P05104.052

Sappers Ross Jelley (left) and Henry Baggaley of 1 Troop, 1 Field Squadron, use a mine detector to clear the way for an armoured personnel carrier (APC) south of Baria. Later, tanks from 1st Armoured Regiment would be offloaded from landing craft and take this road to Nui Dat. AWM neg no. EKN/68/0018/VN P05923.008

A D8 bulldozer, fully armour-plated, builds an earth ramp for loading trucks. Normally the boxed-in area would open. And one reason the armour was added—this D8 (left) has just been hit by an RPG.

The Horseshoe, named after its distinctive shape, provided a natural fortress for a Fire Support Base and home for the Sappers and their infantry and armour protection. The Horseshoe was where the notorious Barrier Minefield was started.

Nui Dat, home of 1 ATF, from the air. The area near the lake on the left is where Sandy's men of 3 Field Troop first established their lines and the task force water point. The flat area just above left of centre is the Kangaroo helicopter pad and the construction of the Luscombe air strip (lower right) required the main road to Baria to be diverted (top of photo).

Sappers Al Stewart, Peter Krause and Dan Brindley set explosive charges to destroy Jumping Jack mines recovered in a day's haul from the mine-clearance operation at the Song Rai river crossing where 17 Construction Squadron built Bridge 6.

An Australian solution to a Vietnamese problem—a typical outback windmill pumps water from a well to a tank, constructed as part of the Civil Affairs program.

This United States Army M-48 Patton tank, crewed by Australian Sappers, was fitted with mine-clearing rollers and lent to 1 ATF to clear possible roadside mines around Nui Dat. Photo by Brian Wood/AWM neg no. P05002.026

M113As of 1 APC Squadron about to leave Nui Dat with troops of 6RAR, for search and destroy operation Ingham. The vehicles are all armed with .50 calibre M2 Browning heavy machine guns with locally manufactured gun shields. Sapper mini-teams were usually mounted on the lead vehicles. Photo by Gabriell Carpay, 1966/AWM neg no. P01404.030

Sapper Danny Brindley of Mungindi, NSW, helps a mate to clear a bunker. They do so quickly in case a bigger Vietcong force is in the area. Photo by Philip John Errington, 1971/AWM neg no. PJE/71/0145/VN

Lance Bombardier David Carswell of the New Zealand artillery sets up his radio equipment to maintain contact with the Kiwi gunners providing support fire for the Australian troops who uncovered the huge Vietcong tunnel complex during Operation Crimp in the Ho Bo Woods north-west of Saigon. A Vietcong tunnel entrance can be seen in the foreground. Photo by Kenneth Ray Blanch/AWM neg no. BLA/66/0026/VN

Australian and American Engineers work side by side to replace a bridge across the Dinh Rach-Hoa river near Baria blown up by Vietcong the night before. Sappers from 17 Construction Squadron, aided by US engineers from Long Binh, completed the twelve-pontoon bridge in 32 hours. Photo by David Reginald Combe/AWM neg no. COM/69/0573/VN

One massive beam is positioned at Bridge 6 on Route 23 crossing the Song Rai river. Bridge 6 was the largest single construction project undertaken by Australian Sappers and helped turn a 'goat track' into a highway.

Lance Corporal Doug Sanderson gets radio operator Private Dudley Fisher to tell HQ he's going back into a tear-gas filled tunnel to bring out an enemy soldier trapped there.

Little Bear—the logo of the Behr-Manning company—was adopted and adapted by 17 Construction Squadron. Amazingly, the Vietcong decided any vehicle bearing this logo was doing good work for the country and was therefore granted safe passage through enemy territory.

A Skycrane tries to deliver a new kitchen to Nui Dat base. The dust storm it kicked up prompted a 'Keystone Kops' search for a suitable landing spot. The load was dumped in the middle of the task force helipad, impeding combat choppers and creating another task for Sappers.

BLACK SATURDAY

'Hammersley'. The idea was for two companies to drive into the Long Hais in APCs, while a third laid ambushes on likely escape routes. More two-man Sapper teams from 3 Troop were attached to the 8 RAR companies, as well as a nine-man combat engineering team from 2 Troop. Anti-personnel mines were still an issue, but because the main thrust would be mounted in APCs, the risk was considerably reduced. Or so they thought.

As C Company reached the Vietcong defences, they came under intense fire from all sides; the lead APC was hit by an RPG and burst into flames, killing the APC commander and driver. This APC was carrying the five-man C Company headquarters team, who were all badly wounded. Despite withering enemy fire, all five wounded were recovered by the combined efforts of Corporal Ronald Macey of 3 Cavalry and Lance Corporal Barry Coe of 8 RAR, under the noses of enemy soldiers who were trying to steal the 50-calibre machine gun mounted on the APC. Both Macey and Cole were awarded Military Medals for their bravery.

C Company withdrew while artillery and air strikes pounded the D445 positions, and when they went back in on 21 February, according to Brian Florence's forthcoming history of 1 Field Squadron Group, the Sappers searched and destroyed more than 80 bunkers containing 35 different kinds of weapons and ammunition, as well as equipment, food, clothing, documents and medical supplies. So far, so good. But with the recent arrival on the ground of A Company, the entire battalion was now concentrated in an area known to be laced with anti-personnel mines—it was an area they had religiously avoided for that very reason.

On 28 February, A Company was ordered to advance to a new ambush area. They were preceded by a splinter team of two Sappers, who swept the route for landmines. They reached their position safely but contacted headquarters to report the discovery of a grenade booby trap. Minutes later, they announced that there had been a 'mine incident' and 17 members of the platoon had been either killed or injured. The official line is that when the Sappers returned from the head of the platoon to destroy the grenade, they missed an M16 booby trap, and one of them, Sapper Terry Binney, stepped on it and was seriously wounded; his offsider, Sapper Ron Hubble, was killed.

The commander of 3 Troop, Captain Peter Thorpe, was doing a recce in a helicopter at the time but was diverted to the scene. On landing, he saw an infantry corporal using the mine-detector; even though Sapper Binney was badly injured, he had told the corporal how to use the device. Binney was subsequently Mentioned in Despatches.

To compound things, when a dust-off helicopter was coming in to pick-up the wounded, the infantry corporal who was using the mine-detector to establish a landing space for other choppers was knocked off balance by the downdraft of the helicopter. The corporal was blown out of the safe lane that he had established and triggered another mine. Even more soldiers were killed and injured, including Sapper Jack Miller, who was being winched down to assist.

Peter Thorpe's two-man Bell 'bubble' chopper then

ferried in Sandy's brother, Corporal Chris MacGregor, who had another mine-detector, and his number two, Sapper Graham Harvey. They had been on the other side of the mountain with another platoon when they heard the explosions. The helicopter delivered the splinter team one by one and landed them on a rock, from where they set to work.

What greeted them was a scene of such terrible carnage that today they either can't or won't remember it. There are some memories best left unstirred—and often your mind doesn't give you the choice. What we do know is that Chris and Graham painstakingly picked their way through the mines, marking safe paths so that the medics could get to the wounded and dead, while Peter took control of the evacuation of the casualties. The deaths of so many Australian soldiers drove a wedge between some Sappers and infantrymen for many years. Had the Sappers been too careless? Had the infantry been too gung-ho? Or was it just a case of a cunning enemy laying traps around traps that were bound to get someone, somehow, sooner or later?

The repercussions of the operation—in which a total of 11 Australian lives were lost and 59 soldiers were wounded—echoed back to Australia. The death toll hit hard in a country that was still generally in favour of Australia's involvement but was beginning to lose its appetite for war; Operation Hammersley was even condemned in parliament. Some say that Hammersley was the beginning of the end of Australia's involvement in South-East Asia. Certainly, the Australian Task Force never sent troops into the Long Hai mountains in such numbers again.

14. CLEARING THE MINEFIELDS

After the Barrier Minefield proved to be such a deadly failure, Australian engineers were instructed to start clearing it. It was 1968, just a year after its construction had begun, against the advice of the Engineers' leadership. Having made lifting the mines as hazardous as possible to try to foil the enemy, Sappers now faced exactly the same dangers, only many times over, if they tried to remove the mines by hand. So a search was launched for a quick, easy and safe method of clearing the mines. Options included everything from armour-plated bulldozers to APCs with steel rollers attached to their sides, like massive metal outrigger canoes.

David Roper remembers the operation well. 'I was posted back to 1 Field Squadron for my second tour in 1969,' he says. 'At this time, the squadron was endeavouring to clear the minefield using APCs that had a long axle attached at 90 degrees to the rear of the machine, which had on it a number of tyres—it resembled a road roller. The idea was to drive around the minefield and detonate the mines.'

This proved unsuccessful because the mines had sunk into the wet soil so not all of them were being detonated. So the next idea was to cut a two-metre-deep trench across the field and strip the top 12 inches of soil, along with the mines it contained, into it. 'After settling in,' David continues, 'I was sent out to the Horseshoe quarry to operate a D8 bulldozer. I'd only been there a short period of time when my dozer and me were transported back to the Dat, where the machine went into the workshop. For the next week I helped the mechanics put steel plating around the engine and cab. Five armour-plated visors from an APC were installed—one front-left, one front-right, one each side and one rear—to give me some vision. A hole was cut in the roof so I could enter and exit.

'After a few days getting used to operating with only the top-left corner of the blade visible and very limited vision elsewhere, I was sent out to the minefield with my D8. I then started a procedure that was to become my regular routine. The Vietcong were becoming upset at losing their supply of mines, so they started to booby-trap the wire and mine the road. I would lightly skim the road surface on the way out, and where the wire was to be cut for access I would brush it with the machine. This proved to be successful as I would quite often detonate booby traps. I think the worst one was a Claymore—possibly because it was the first. It left a lot of scarring on the cab plates.'

Working in the field was a much bigger challenge. The idea was to get a good blade full of soil and then, using that as a cushion against any mines that detonated, scrape the top foot of soil into the trench. 'The first mine I set off sent the gauges out of whack,' says David, 'although I sometimes think it could have been that the shockwave put my eyes out of focus and filled the cab with descending soil. I never did get over the fear, but after the first couple of days I learned to live with it. The Vietcong started sniping at me so a Centurion tank was sent out to be my protector. He would set up outside the wire, and when I was taking incoming fire I would try to give the tank crew the coordinates over the radio so they could return fire. This usually made the enemy lose interest.'

Ventry Bowden was another plant operator involved in clearing the Barrier Minefield. 'The powers that be came up with all sorts of weird ideas,' he says, 'but the best plan was to use an APC, fitted with a large length of water pipe, tightly packed with old truck tyres, and on the end of the pipe a chain was welded to the front of the APC. The vehicle would enter the minefield and drive around in ever-diminishing squares, with the truck tyres exploding many of the mines. It still left many more clearly visible—they were lying at all angles on the ground.

'The bulldozer, which was fully armoured up, with metal plating all around it, would also enter the minefield. It would crawl along one side, making sure to keep very close to the barbed wire, and then would gouge out a very deep trench, at the far end of where the APC had partially cleared with its tyres! This trench was dug from one side of the field to the other, wire to wire, and when it was considered deep enough, the dozer would reverse right up to where it first entered the field. Then it would start pushing layers of dirt, along with unexploded mines, into the trench.

'I accidentally set off quite a few mines by reversing over them unseen. The dozers had very small windows mounted at each side of the operator's cabin, to enable the driver to see how deep he was digging his blade and where he was going. There was also a small window mounted at the rear of the cabin, for reversing. These small windows were similar to the ones installed in tanks, as they were very thick and could withstand the exploding shrapnel. On different occasions, the bulldozer would have a track blown off from an explosion. You could certainly see how much devastation an M16 mine would cause a human being, if they could cause so much damage to a 50-tonne machine.'

Typically for Sappers, despite the potentially deadly work, there was no shortage of amusing incidents. David Roper recalls trundling back into camp one day, only to see everyone diving for cover as he drove past. He couldn't work out what the panic was until he realised he'd come back with an unexploded anti-personnel mine sitting in the radiator grille. 'To remove it, I had to lie on the bonnet with a piece of cord and lasso it,' says David, 'which, after a number of tries, I did, then I lowered it to the ground and exploded it.'

The Barrier wasn't the only minefield that needed clearing. Sometimes the enemy had laid their own traps (often using recycled Allied munitions) that had to be dealt with, as Allan 'Blue' Rantall recalls. 'Christmas 1967 was good, and I was looking forward to New Year's Eve, when that morning four of us got ready-reacted out bush,' he says. 'Apparently, the SAS had come across a number of M16 mines scattered on the floor of the jungle. We were choppered out and winched into the area with our mine-detectors. It was all so crazy—none of us had ever been winched into anything before. I don't know who was controlling it all, but we were told to run for a couple of hundred yards before we set up the kit.

'There were no grunts with us and we didn't see anyone else, so I can only assume that the SAS were in the background somewhere. We located about half a dozen mines, neutralised them and then we were winched out. We got back to Nui Dat in time for the celebrations.'

Another reason for clearing minefields was simply a change in tactical priorities. It was standard practice when setting up a defensive position to establish a minefield, to limit the options the enemy had if they wanted to attack an installation. The Horseshoe was a good example of a minefield that was clearly marked; it wasn't there to catch an unwary enemy but was intended to deter them from approaching in the first place.

However, as time went on, a defensive minefield might be in the way of a new strategic installation, and would have to be cleared. Peter Krause found himself unexpectedly involved in just such an operation. He was clearing a minefield that had been laid by the South Vietnamese Army to protect the

CLEARING THE MINEFIELDS

approaches to a bridge on Route 23 crossing the Song Rai river. How effective it would have been was debatable, since many of the landmines that were removed still had their transportation safety plugs attached. But clearing this minefield was exactly what Peter didn't want to do.

'Even though it is 40 years ago, I can still remember the day at the School of Military Engineering when they demonstrated the effectiveness of an M16 mine, amongst other things, on the range,' he says. 'Our training troop observed from 100 metres away. I remember the puff of smoke and the pop out of the ground—you have just got time to think, "This is not too bad"—then you observe the red ball of fire when a pound of TNT encased in eight pounds of cast iron explodes a metre off the ground with a crack that I have never forgotten. Like everyone there, I wondered what demented bastard had invented such a thing, and I decided there and then that I would avoid ever having anything to do with M16 mines.

'Almost twelve months later to the day, while serving in 1 Troop, 1 Field Squadron, I was deployed as the number one of a mini-team supporting the mine-clearing team. The group contained two D8 dozers (armour-plated by 1 Field Squadron) with Corporal Gordon Daley, and Sappers Des 'Monto' Curtis, Wayne 'Jess' Jessop, Les Dennis and Peter Haig. There was also one APC with a driver and a medic, and Sappers Al Stewart, Dan Brindley and myself from 1 Troop.'

Peter recalls that the armour-plated D8s were the most effective mine-clearing machines in a long line of experiments that enjoyed limited success, but all the plant operators had to see through was a couple of square inches of armoured glass. 'They had no means of communications, and if we wanted them to stop we would drive to them in a closed-down APC and throw a smoke grenade. They would stop and emerge out of the top of the old D8 to find out what we wanted.'

For most of February 1971, the team cleared the minefield surrounding the bridge, and once again the D8s bulldozed out long, deep slots, strategically placed in the minefield, and then bladed across the field, pushing dirt, mines and other rubbish into the slot. When finished, they pushed the 'borrowed' dirt over everything to close the trench. And that was the end of the mines and the minefield—in theory, at least.

'This was where the mini-team came in,' says Peter. 'Our role was to do a visual over the dozed earth. We had to recover and destroy the mines we found—which turned out to be quite a lot. They were in various states of disrepair—some were cut clean in half, while others were primed and ready to explode. We found that a lot of mines had spilled out from the edge of the D8's blade, so we dragged an old Bailey-Bridge Panel with an APC to expose the mines that were left. We recovered up to 60 a day, and about 600 in the whole field.'

Peter didn't have any experience with in-country mine-clearing, so he and his number two, Al Stewart, had to improvise. 'We had only a few locking pins when we started, and we used a rusty old shifting spanner as an arming tool. Later we found some arming spanners in the minefield and Al made a stack of mine pins from wire off the ration packs with a pair of pliers and a sharpening stone. Sappernuity! We did everything we could to recover as many mines as possible—we were very aware of the way our own mines had been lifted in great numbers and used so effectively against us. We screwed out the igniters, broke the flash caps and pulled the trip-wire loop, setting off the fuses and making them unserviceable. There were areas where, for whatever reason, the mines were deeper in the ground—probably the very wet areas—and when we suspected this, we resorted to using a mine-detector. We found quite a few mines this way too.

'We destroyed the recovered mines daily, packing them in a pyramid pile with a couple of sticks of C4 on top,' says Peter. 'Kaboom! Some spectacular explosions. On one occasion, the top of a blown-up mine came down through the thatched roof of one of the nearby village houses. It was so hot that it set fire to the thatched roof and landed on a century-old crockery teapot, breaking it. The old bloke—nearly as old as the teapot—picked up the red-hot mine top with his bare hands, burning them badly. My colleagues pointed to me when he came to see us for revenge and compensation—that was big of them!

'Some medical attention, rations and cigarettes was all that we had to compensate him with—and this disappointed the old bloke no end. The heavy crockery teapot seemed to be the greatest loss to this old guy. I was concerned that it was an irreplaceable part of his family's meagre possessions, and I wondered how long it would be before they could replace it.'

Injuries and embarrassments from munitions blown up for fun could come loosely under the category 'self-inflicted wounds' but in terms of the amount of low-level damage done to the greatest number of Australian soldiers, Sappers included, nothing could compare to the myriad charms and temptations of Vung Tau, the Diggers' R&R haven in the south.

15. SIN CITY

It's fair to say that *Tunnel Rats* broke new ground in military histories by revealing what soldiers did when they weren't shooting at the enemy and being shot at—or, in the case of the Sappers, exploring tunnels and digging up mines. Mostly for the bad boys of 3 Field Troop, it was laying siege to the bars and brothels of Bien Hoa and Saigon, activities which may these days seem inappropriate, especially today when attitudes to sex and women in general have moved on considerably—as they needed to.

But the reality in Vietnam during the 1960s was very different. On the one hand, you had a bunch of young men in their physical prime who truly didn't know if their next day might be their last. On the other, you had an enthusiastic local trade in the kind of distractions that would take their minds off their mortality... if only for a night or two. And the place to go for those often forbidden pleasures was the once elegant seaside resort of Vung Tau, established as their version of Cannes or Nice when the French ruled Indochina.

To be fair, for many young lads, R&R in 'Vungers' was no worse than a buck's night, although with the excuse of imminent danger rather than impending matrimony. There was simply lots of drinking and messing about with mates. A few punch-ups and letting off steam, some bullshit and bravado, but nothing you wouldn't tell your mum about. For others, it was a garden of previously unimagined earthly delights, as *Holdfast* editor Jim Marett recalls.

'A trip to Vung Tau was a rare treat indeed for the Tunnel Rats of 1 Field Squadron. Usually, the trip was for two days' leave after a four- or six-week operation out bush. It was always a period of comradeship and drinking, and some erotic pleasures could also be looked forward to over the two days. Not everyone indulged in this latter activity, but for those who did, there were many options available, and all at affordable prices.'

The trip to Vung Tao from Nui Dat took little more than an hour, but it was like entering a different world—not just different from army camp life, but different from anything any of the men had experienced back in Australia. And it wasn't just the sex and the booze. 'Compared to being out on operations, we felt totally safe for the next few days,' Jim explains. 'All those pleasures were waiting for you, and there was nobody to stop you enjoying them... The comradeship and the sense of urgency that war brought to the package were essential elements in the mad mix that was Vung Tau.'

When it was under French rule, the southern city may have been Vietnam's equivalent of the Côte d'Azur but there were few airs and graces when the Aussies came calling. The first stop for accommodation was the Peter Badcoe Club, an army-run facility with real beds, proper food, a bar and a swimming pool—which, with typical Aussie gallows humour, was named the Harold Holt Memorial Swimming Pool.

'At check-in, we'd hand in our weapons and receive a lecture from the Padre about the dangers the hookers in town were to our health and morals. "Put two condoms on, then stand at the door and watch your mate do it," was the sage advice from one,' recalls Jim. Changing into civvies added to the men's sense of freedom and—let's face it—reckless abandon.

'The bar at the Badcoe Club seemed to be open all the time, so a few beers were the first order of the day,' says Jim. 'And perhaps lunch, too, because you'd probably eat very little over the next few days. There were restaurants and food stalls in town, but we were wary of eating anything that would ruin your hard-earned break with a dramatic dose of the trots. The bread rolls stuffed with cold meats sold on the streets were known as "heppo rolls". There was a simple solution: drink but don't eat.'

The main mode of transport around the town were 'Lambros'—Lambretta scooter-powered passenger tricycles with the same DNA as the tuk-tuks of Bangkok. 'Weighty Aussies soon discovered that by standing on the back step of the cabin you could tilt the whole machine and have the front wheel and the driver in mid-air,' says Jim. A less common vehicle—and pretty much a collectors' item now—were the 'motor-cyclos', deadly machines that had the passengers sit at the front of the vehicle, while the pedal power of the usual cycle was replaced by a small motorbike at the rear. The experience was not unlike being driven around on the bonnet of a small car, with all the obvious dangers that entailed.

Having made it to town, the first priority was to change Australian dollars to Vietnamese dong, which was usually done in an Indian tailor's shop. The transaction was heavily biased in the moneychanger's favour and was highly illegal, but the diggers ended up with reassuring wads of Vietnamese cash. 'The downside was that you'd really be in trouble if you got caught,' says Jim. 'But, hey, what were they going to do? Send us to Vietnam?'

Cashed-up and ready to rock and roll, the guys would usually meet up at the Flags, a physical embodiment of the United States' 'many flags' campaign, where the banners of all the countries supporting South Vietnam were on display. The visitors, however, were mostly Australian and American. Wherever they came from, they had two things on their minds—grog and girls, and not necessarily in that order.

The transactions that were on offer in the various girly bars ranged from buying the attentive ladies overpriced drinks while they draped themselves around you, to going the full 'boom-boom' for a negotiated fee. Bars existed where you had no sooner parked yourself on a stool before delicate hands would be fluttering around your zipper and, somewhere under the table, 'num-num' would be administered. Thus, Aussies could indulge in one of the great male fantasies—enjoying a beer, a yarn with your mates and a blowjob, all at the same time.

However, more often the action was a lot less perfunctory, and there would be a long and slow seduction involving the purchase of large quantities of Saigon Tea—an overpriced concoction that may have been exactly what it said it was—while the girls told their new best friends how handsome, funny, strong and sexy they were.

Even at its most innocent level, the scene was a culture shock for many Aussie recruits, especially those innocents abroad who were encountering the temptations of the flesh for the first time. Trevor Shelley was in Vung Tau almost by accident, while en route to serious action as a number two in a splinter team attached to infantry operating out of Nui Dat. Despite an arduous time getting to Vung Tau, he wasted no time once he arrived. 'Having a fairly good constitution, I managed to get to the China Doll Bar that night,' he recalls. 'I was very impressed that all the girls were in see-through negligees.' Having spent his early life in south-western New South Wales, that's the kind of memory four decades clearly can't erase.

Of course, the contents of the Saigon Tea bore no relation to the prices being charged. 'What you were actually doing was buying a girl's time, paying for her to sit with you,' Jim says. 'Her goal was to seduce you into spending the night with her. In most cases, this was not a difficult task and usually required nothing more from her than a straying hand and maybe a romantic line. "You like boom-boom me?" was always a winner.'

All the bars were different in subtle ways, and the Aussies did import some of their social practices from home—not least the gender divisions employed at a million Saturday night dances and backyard barbecues. Sandy MacGregor vividly recalls Rani's Bar, which was near the Sappers' base at Vung Tau. Rani, the owner and 'mama-san', was Indian and had been well trained by the original Tunnel Rats of 3 Field Troop, back in 1965.

'We were known as "Cheap Charlies", recalls Sandy, 'because we didn't like to buy the Saigon Teas when we were drinking and yarning with our mates. In fact, in typical Australian style, we didn't want to be bothered by girls when we were drinking. So it was suggested to Rani that she should keep the girls away from the guys unless the men invited them over.

'As you can imagine, she took a bit of convincing but it worked. The guys would sit at one end of the bar, laughing and joking and spinning tall tales—and the girls sat at the other end, waiting to be called over. At first Rani couldn't see how she could make any money if she wasn't selling Saigon Tea. But she was clever—she just charged more for the beer and provided beds out the back that the guys and the girls could go to for a quickie, and the bar flourished.'

Rani's Bar was still operating the same way in 1970, with

Rani still the mama-san. Her bar was frequented by Sandy's brother Chris and many, many other Australians.

Those more in search of romance rather than a quick root between beers could take their girls out of the bar for the night. Of course, that involved extra fees, which had to be negotiated with the mama-san and then split between her and the girl.

'Most of the girls lived in little terrace houses in the back streets of town, sharing with other hookers,' says Jim Marett. 'The bedroom often housed two, three or even four double beds, with mosquito nets providing the only element of privacy. This didn't seem to matter at the time, and in fact it provided an opportunity to meet new people. A few US servicemen were usually on leave at the same time as us.

'Bathrooms were non-existent. Instead, there was usually a large earthen jar in an adjoining outside courtyard. Using a wooden dipper, you'd be given a splash bath that sobered you up a little before you retired to the communal love nest.'

The men would usually return to the Peter Badcoe Club for a sleep, a shower and running repairs before heading out for another day's recreation—there was never much rest. Few diggers probably knew that their accommodation was named after the late Major Peter Badcoe, known as 'the Galloping Major'. A member of the Team (AATTV), he won the Victoria Cross in February 1967 when he rescued a US Army medical advisor under heavy fire in Huong Tra and led his company in an attack that turned what seemed certain to be a defeat into a victory. He was killed the following month, aged 33, while attempting a similar feat elsewhere.

The *Australian Dictionary of Biography* describes him thus: 'Short, round and stocky, with horn-rimmed spectacles, Badcoe did not look a hero. He was a quiet, gentle and retiring man, with a dry sense of humour. His wife was his confidante. Badcoe neither drank alcohol nor smoked; bored by boisterous mess activities, he preferred the company of a book on military history.' It's hugely ironic, then, that any diggers unaware of his heroics, associated Badcoe's name with drinking, smoking and playing up. And they would have found it highly amusing that Major Badcoe had changed his name from 'Badcock' in 1961.

During the daylight hours, there was plenty to see in Vung Tau for those who could prise themselves away from the bars. And if the need for some TLC presented itself before darkness fell, there was always the hairdresser's shop. Back in the day, these hairdressers did a whole lot more than short back and sides. 'Manicures, facials, shoulder massages and ear-cleaning were all on offer,' says Jim. 'Even the basic haircut and shave involved shaving the entire face, including the forehead, nose and the surface of the ears. Nose hairs were delicately clipped with needle-like scissors. By the time you left the chair, you were literally shining.'

However, some hairdressers provided an altogether more intimate experience. 'These places did all of the above but also offered steam baths and massages, plus extras,' explains Jim. 'This was a great way to refresh yourself halfway through a day of drinking with the lads. The steam bath was communal—it was great fun with a bunch of mates. The massage was performed by a pretty young girl whose role was not unlike that of a teasing mare in horse breeding. Towards the end of the massage her hand would inadvertently brush against your most sensitive areas while she did her sales pitch.'

After the fees for extras were negotiated, the young girl would disappear and an older and less attractive 'turkey woman'—a gobbler—would take her place to fulfil the contract. 'The romance of the moment was often reduced by one or more of your mates peering and jeering over the low partitions between each massage cubicle,' says Jim.

There was the entire spectrum of entertainments in Vung Tau, including high-class establishments like the Jade Bar—a sophisticated brothel—and the Grand Hotel, which had cocktail bars, gambling and restaurants; it was mainly for officers.

Not all the troops in Vung Tao were there on R&R—there was a permanent military presence too. The Australian Logistical Support Group was there, and 17 Construction also had a base, since the port was the main arrival and departure point for men and machines on their way to and from Australia. Even so, their lives were very different to those of their comrades operating out of Nui Dat.

When Corporal 'Stiffy' Carroll was assigned to Vung Tau, prior to being shipped back to Australia, it was to a troop responsible for water supply to the Australian Task Force, which was expanding rapidly. Stiffy was put in charge of the water point—massive storage and purification tanks— that had been established by 3 Field Troop under Sandy MacGregor. The water point was a considerable achievement in itself, with Sandy's men digging out a pond in the sand dunes, working against a constant in-rush of water to get as deep as they could so as to provide the vast amounts of water that would be required by the troops that were soon to arrive from Australia. Sandy then incurred the wrath of some senior officers by commandeering giant rubber 'pillow' tanks, intended for fuel supplies, to store treated water. He is rightly proud of the hard work and 'sappernuity' that went into creating the water point at Vung Tau.

'Our living quarters consisted of a marquee without walls on top of a ridge of sand next to the storage tanks, with eight fold-up stretchers for the workers,' says Stiffy. 'All meals were delivered to the water point—as was our pay, too—because we were a long way from camp. Meals were generally eaten sitting atop the pillow tank. Nobody bothered us. We were virtually a law unto ourselves.

'It wasn't long before word got around that our water was far superior to that of any other supplier. The Yanks mostly sterilised their water with iodine, and so it smelled like a hospital's urinal. We began to get requests for sly water from several Yank units. I've always been one for furthering international relations, so using a little diplomacy, I explained that I could get into strife for breaking rules; worse still, my men would have to work beyond normal hours.

'You wouldn't believe how generous and understanding our friends from across the Pacific could be. As soon as darkness fell and our own trucks stopped, the Yanks would arrive. Our biggest customer was a US engineer named Ray Leibske, and he always left a couple of bottles of bourbon or other commodities that we needed. Some of our customers would leave cartons of beer or spirits, or bags of ice and soft drinks.'

Stiffy and his crew converted an old Second World War Patterson water trailer into a mobile canteen, using donated ice and locally purchased mud crabs. 'It became a mobile morale booster,' claims Stiffy. 'You'd be amazed at who dropped in to get instruction on how to operate a water trailer. We had also acquired a couple of bomb crates, alloy cylinders lined with polyfoam, which, when buried end down in the sand, became the best ice box you could ever wish for. The night shift became quite a social event.'

To keep a good flow of water going in the well, Stiffy decided to pump straight from the well into a depression in the sand dunes about 50 metres behind the water point. The idea was that the increased inflow from the water table would keep the water fresher and prevent silting. It also created a small lake, and the discharge from the four-inch pipe made a great massage shower.

'The queue for this luxury—plus a swim afterwards—was remarkable,' says Stiffy. 'I might add that there was an entry fee. I was as happy as Larry. No one was shooting at me, I had plenty of booze, I didn't have to listen to anyone, I was my own boss. Some of the characters who worked with me there were Lennie Higgerson, Lance Guthrie, Andy Anderson, Tommy Elder, Nipper Simpson and "General" North.

'Andy Anderson would sit in a dinghy on the water on his days off with a fishing rod and reel, and wait for passers-by to ask what he was doing. He would yell, "Wait a minute—I've got one!" and go through the motions of hauling in a marlin. At that point he'd pull out a can of sardines attached to his line! Looking disappointed, he'd declare, "Oh, it's not as big as the other four!"'

Stiffy was happy to share his good fortune—for a fee—but there were always those, especially in the officer classes, who thought they were entitled to whatever was going around. 'There was the sergeant who turned up at midnight with a water trailer behind his Land Rover, demanding that we fill him up for free,' recalls Stiffy. 'He said the water was for his commanding officer's private water tank, which had just run empty. He suggested that if we didn't comply, large ripples would result. Lennie and I filled his tank straight from the well, and to purify it we added a bucket of chlorine and two buckets of alum.

'A couple of days later, I had a visit from the health inspector. He was anxious to know what might cause milk in a cup of tea to go solid and sink to the bottom of the cup, and why a certain officer, after showering, had found that his comb got stuck in his hair. Of course, I didn't have a clue—there were no records in our logbook of us ever having supplied his private tank with water. Not long after that I was flown home.'

Vung Tau wasn't all high-jinks and low-life bars. There was plenty in 'Vungers' for those who wanted to 'keep themselves nice', but either way, sooner or later diggers had to pack their gear at the Peter Badcoe Club, collect their weapons and get on a convoy back to Nui Dat. And for those who'd overdone

SIN CITY

their farewell celebrations there was a final sting in the tail. Just outside town, there was a fish sauce factory. Fish sauce is made from fermenting anchovies, so you can imagine the smell around the place, wafting into the trucks and buses transporting soldiers, many of whom were already pretty green around the gills.

Legend has it that when the Americans accidentally bombed a fish sauce factory in the Mekong Delta, the cloud of 'fall-out' was so intense that they thought they had blown up a chemical weapons factory. For the troops, hungover and 'rested' to the point of exhaustion, bouncing along rutted roads and assailed by the aroma of rotting fish, a deadly nerve gas attack might have come as a blessed relief.

16. THE TETHERED GOAT

The secret of a trap—any kind of trap—is that the intended victim must see it as an opportunity rather than a danger. So when it became obvious during the latter half of 1970 that the Vietcong were skirting Phouc Tuy via Long Kahn, the province to the north, the Australian Task Force commanders decided the best way to cut their east-west supply lines was to lure them south and entice them into an area where they could be more easily attacked. But what should be the bait? A soft target such as a road construction unit, perhaps? Not so well defended that the Vietcong gave it a wide berth, but not so obviously vulnerable as to make the enemy suspicious.

That's why, when the ATF got the okay from their American Allies to patrol and ambush four kilometres into Long Kahn, they decided that improvement of the Route 2 road provided a credible reason for being there. A night defensive position (NDP) to protect the construction squad would be established near the village of Cam My—which would make perfect sense to the enemy—but infantry from the NDP could also launch raids into Long Khan, drawing even more Vietcong into the trap.

The Vietcong didn't realise that the construction squad at NDP Garth was a 'tethered goat'—but, then, neither did its officer in command. 'We were building Route 2 from Baria up to the border with Long Kahn province, which is to the north of Phuoc Tuy, and the road had reached Bin Bah,' says Lieutenant Roger Cooke. 'The Australians needed an excuse to get a bit of land off the Americans so they could put us in there as the "tethered goat"—in other words, to try to persuade the Vietcong to have a go at us, then hit them with superior firepower. I didn't bloody realise this until I read Paul Ham's book, *Vietnam: The Australian War*. I was stupid enough to think that I was doing a real job—we were up there building a road thinking we were doing a great thing for the country.'

The initial development of NDP Garth was carried out by 1 Field Squadron's Plant Troop. But road-building fell under the remit of 17 Construction Squadron, so when Roger Cooke arrived in October 1970 he found that his duties changed overnight. 'I was originally in 1 Field Squadron when I arrived,' he says. 'I don't know how they had got themselves a road construction troop because they were supposed to be field engineers and combat engineers—the construction troop was supposed to build the roads. Anyway, I was with 1 Field for about two days, then I was transferred to 17 Construction Squad and took over NDP Garth.'

The roadhead was too far away to be economically and safely constructed from the Task Force's base at Nui Dat. However, a source of road base material was necessary and, fortunately, a potential quarry site was found in a hill near Cam My. Everything was falling into place. The Australians had a valid reason for being there (other than wanting to engage the enemy), they made a tempting target, and they had the materials required to make this a practical proposition. All that remained was to negotiate payment to the land owner for access to extract the rock, and bring in the necessary equipment, including a rock-crusher.

'The most likely source of material was a volcanic plug at the western end of the airstrip that serviced the De Courtney rubber plantation at Cam My,' says Roger. 'Fortunately, this was adjacent to NDP Garth. A family in Cam My had coffee trees on the hill and so the Department of Civil Affairs negotiated compensation. This is when it became obvious to me that the South Vietnamese men were the head of the families but the women had the brains. The husband was ready to settle for anything but the wife refused to sign and pushed for a very good price.

'An access track was cut to the top of the hill, where an area was cleared for a standing patrol and a listening post to be established. Then the topsoil was removed, revealing scoria. We were hoping for basalt or granite but scoria—which is a porous volcanic rock, better suited to landscaping and barbecues than road building—was all we got. Once we realised that, a bit of Sappernuity was applied,' says Roger. 'The D8 was to raise the scoria on the hillside and push it down to a bench at the bottom of the hill, where the TD15—a smaller bulldozer—would push the material on the flat to a chute that fed into the crusher.'

Construction squadrons weren't usually expected to be in combat areas, Roger explains, so the first priority was security and defence. Wire was laid, Claymores were positioned, the command post was dug in and four strong points were constructed at the corners. The infantry and engineers built individual firing positions from ammo boxes filled with soil (four boxes high) and four-foot half-galvanised culvert pipes (to make a roof for undercover sleeping), as they awaited the construction of a main embankment or 'bund'. A D8 bulldozer was used to push up a six-foot bund between the strong points. The ends of the firing positions were then cleared of soil and stabilised with sandbags. Roger Cooke used the burnt-out hulk of an American APC for his hootch.

'They put a 105-mm howitzer there, although it was very unusual for the artillery to split up a section of guns because they were supposed to be mutually supporting,' explains Roger. 'They had one howitzer and one mortar, but the intention was to use them mainly for illumination because we actively patrolled and ambushed around the perimeter. Anytime we made a contact we put up so many lights that no one bloody moved.' Artillery missions often included firing parachute flares (illuminations, or 'illums') to provide light for ground forces at night. Having sole howitzers and mortar tubes went against all the usual rules of deployment but it would have added to the impression that Garth was a soft target. Companies of 2 RAR were rotated through the NDP, as were sections of APCs to protect both it and the roadworks. Roger was promoted to captain so he could deal with infantry and armour troop commanders on an equal footing.

'Well, I was the one who was there all the time,' Roger says. 'Every now and then a major would come in so I'd be out-ranked, but the rest of the time I had the power of a brigadier to say, "No, you're here to protect us, not to piss off and chase the enemy."' Roger nevertheless soon had his men doing a lot more than road-building. 'The infantry was there for our protection, and the cavalry had a section of APCs, but they were always under strength so I allowed some of my guys to go out on ambush—although not every night—and we'd man the guns on the strong points at each corner. There were construction squadron personnel on an ambush one night that snared 14 Vietcong.'

With different units rotating through Garth, Roger encountered the entire spectrum of Anzac infantry. 'The Kiwi battalion had a lot of Maori in it, and I remember one bugger, he thought he was a tank himself. He carried the M60 machine gun, I was coming back from an ambush and he arrived on their APC, and he stood up on the back of the APC—carrying the M60 and all his frontline ammunition—and just stepped off it and just walked into the bush. I used to crawl down the side of the bloody thing, keeping my head down.

THE TETHERED GOAT

'They were terrific soldiers, but their pride in their toughness sometimes came out in doing some terrible things to bodies. On the other hand, we got a Vietcong one night and I was taking a photograph of his body, and a Maori corporal said to me, "Would you like them to be taking photographs of you if you were dead?" That really struck home to me and I never took another photograph of a dead body again.'

It was an odd situation for the squadron, doing construction work during the day and patrolling the surrounding area at night, but it wasn't without its distractions. 'After a certain hour every night there was a curfew when all the locals had to be in their villages,' Roger says, 'but during the night we could hear this bloody Lambretta scooter roaring up the road through the rubber. We'd go up that road the next night and he would roar up behind us or in front of us—he was obviously planting mines.

'One night I was going out on ambush with the APCs and we struck a mine. I was in the bloody vehicle and I stuffed my knees on it. Thankfully, they had armoured the bottom of the APCs with an inch of aluminium plate—that saved us a lot. Anyway, I had a land-clearing tank so I got it to come out and clear around the APCs that had been damaged. The tanks started towing the APC away, and all of a sudden I hear *brrrrrrr*... The Lambretta driver had obviously set the first mine off, and there must have been another one around—he was waiting around to see us hit it. I'd love to meet the guy—I hope we never got him.'

Military manoeuvres aside, the main reason for them

being there was—officially, at least—to build the road. November was too wet to work on the road, Roger recalls. The time was spent sourcing a rock-crusher. A brand-new mobile Eagle crusher, with accompanying conveyor belt and generator, was bartered from the Americans at Long Bien, reportedly for four slabs of VB and four pairs of Australian Army boots. Once it was installed and working to Roger's satisfaction, it was time to have a look at the road that was to be upgraded.

Roger took Sergeant Ken Sneddon and Corporal Bernie Gilles on a recce of the area to determine where they would need culverts and of what size. This scouting exercise nearly ended in disaster, which was only averted when Roger looked down and—purely by chance—spotted a trip-wire. It had been set up by 'friendlies' of the Regional Force, based at Cam My. Needless to say, the locals were persuaded to delouse all similar traps in the area.

In December the ground had dried sufficiently for work to start. The commander of the infantry section protecting the immediate area told his gun group to set up in a green thicket beside the road. They stepped into it and promptly disappeared. It turned out that the undergrowth covered a dry well, and the crane on the fitter's track was needed to get the two soldiers out. They reported that there was a mongoose stuck down the well too, so Craftsman Jones went down and rescued it. The mongoose became the detachment's mascot and was named Rikki-Tikki-Tavi.

When they finally got to work, the first cut by grader unearthed a home-made mine; luckily, the grader's metal blade had cut the wires but not completed the circuit. It was a standard home-made anti-tank mine of about 20 kilograms of Chicom explosive. 'It would have made a real mess of the grader and operator if it had gone off,' says Roger. 'Further roadworks were delayed until four mine-detectors could be delivered. Clearance using mine-detectors was ridiculously slow—they were detecting so much shrapnel and vehicle parts that something else had to be employed that would clear at least 50 metres of both shoulders each morning.'

The answer came in the form of 1 Field Squadron's mine-clearing tank—an American M-60 with multiple rollers on an extension in front of it. It was effective, but every time it hit a mine a whole day was spent reinstalling the rollers. Also, the Vietcong sometimes placed the pressure switch beyond the explosive, so the mine would go off under the tank rather than the rollers. 'The tank was a moulded hull so the blast didn't damage the inside,' says Roger. 'But it also acted like a giant bell—the operators' eardrums took a hammering.'

It seemed that the roadworks would take many months and NDP Garth would be 'home' at any given time for an infantry company headquarters, at least one platoon, a section of APCs, two mini-teams from 1 Field Squadron and 30 plant operators from 17 Construction. Recognising the need for a few home comforts, two Sappers, under the direction of David Mumford, constructed a toilet and shower block for the use of all, and a mess hut called the Cuttamundra Pub for the engineers. David Mumford—affectionately referred to as

'Mum'—prepared bacon and eggs every morning in the pub so the engineers were always ready to go at daybreak.

Once everything got ticking along on the roadworks, 40 metres to 50 metres were 'boxed out' each day, with sub-base and base courses brought from the quarry, spread and compacted. No more was prepared than could be constructed in the day, to avoid mines being put in the newly disturbed subgrade. Every now and then things would go wrong, however, so more was boxed out than was filled and compacted.

In January 1971, an Acco tip truck was travelling along an area that hadn't been finished and hit a mine that obviously had been laid the night before. The mine went off on the passenger side—only the fact that the engine in an Australian-built Acco was between the passenger's and driver's side shielded the driver, who simply had burst eardrums. Chicom explosive gives off very black smoke so the driver was covered in black soot; when Roger got to the truck he called the driver Vic because he thought it was an Aboriginal Sapper, Vic Slockie. In fact, Vic was operating the D8 dozer in the quarry.

The cavalry APCs spent their days protecting the roadhead and their nights ambushing around the area, hoping to intercept Vietcong intent on mining the roadworks. One day, the cavalry sergeant Paddy Waughe noticed four women walking along the road, evenly spaced like an inline patrol formation. He gathered them in for interrogation. They were, in fact, an 82-mm mortar crew who had already set up their tube and base plate with four rounds just north of Cam My. They had been planning to mortar NDP Garth that night.

THE TETHERED GOAT

Another ambush in April 1971 killed a Vietcong courier who had documents telling the local VC not to mine the roadworks; they reckoned they would eventually win the war and that the road would be an asset. 'The worry was that we had literally shot the messenger who was carrying this important order,' says Roger. 'We really would have preferred him to get through, but there must have been other couriers who got through because the mining of the road stopped.'

However, the hands-off directive did not apply to APCs in the rubber plantations. 'Ridiculously, there was a political directive that the APCs could not cut across the rubber plantations because their aerials would scar the rubber trees. It didn't take the Vietcong long to realise this, and they confined their mining to the tracks through the rubber. There were six APCs that hit mines around Garth.'

Rubber was still big business in South Vietnam, and Roger Cooke got to know Patrick, the French manager of the De Courtney rubber plantation, who asked him to lunch at the Plantation House. Saigon was still an open port to tourists, so Roger invited two young women friends to call in on their way back to Australia from London. Thus, Captain Bruce Hughs, Lieutenant Bill Carlton, Roger and the two young women, Gini Scott and Rose Cullen, had lunch right in the middle of Vietcong territory. Roger found out years later that the SAS had Patrick under observation—and even recorded the lunch—because it was suspected that his bosses in France gave him money to pass on to the Vietcong so they wouldn't interfere with the operations of the plantation. In fact, Patrick was later killed by the Vietcong when he visited another manager near Xuan Loc who had not been paying the protection money—he was pocketing it. A Vietcong hit squad blew up his office, killing Patrick as well.

Meanwhile Roger had made a bet with Deputy Task Force Commander Col Forward—for a bottle of champagne—that basalt would be found at the quarry. As it turned out, Col was correct, except for one basalt 'floater' about a metre in diameter. Roger had the rock put in the back of one of the Acco trucks and told the driver to drop it at Col's hootch in Nui Dat.

'Unfortunately,' says Roger, 'the driver was not told of the bet and thought he was doing something wrong. So he dropped the rock at the colonel's door and made a run for it with the hoist still up, taking all the overhead telephone lines for a hundred metres or so with him. It took days to restore the communications in Nui Dat.'

The commanding officer of 17 Construction was Major John Sanderson, who later became the chief of general staff and, later still, the governor of Western Australia. He was very supportive of the detachment at NDP Garth, recognising its true purpose. He also realised that the Task Force needed the pretence of the roadworks. For some reason, however, the 2 RAR regimental medical officer reported to the battalion commander, Lieutenant Col Church, that hygiene at Garth was substandard and that Roger Cooke was responsible. The battalion commander charged Roger, which infuriated Major Sanderson because he fully realised all that Roger had done at Garth to ensure that the hygiene standard was high.

THE TETHERED GOAT

'Major Sanderson went in to bat for me with such fury that the lieutenant colonel withdrew the charges,' says Roger. 'The good major then took the matter further and immediately stopped work on the road and withdrew the 17 Construction detachment, with all the plant including the crusher. Unfortunately, 100 metres of road past Garth and Cam My had been boxed out ready to be filled with crushed rock that was sitting in the quarry, but the road had to be left unfinished.'

Soon after, the Task Force's Operation Overlord activated the trap to catch the Vietcong who had been attracted into the area. The now-abandoned quarry was used as a base for the tanks, and a total of 137 bunkers and one tunnel system were located and destroyed in a five-day operation. Roger Cooke served a total of 14 months in Vietnam, including time with the US Rangers, which saw him more actively involved than he had intended.

'I was seconded to the American MacV compound in Xuan Loc to tell them what they were doing in our province,' recalls Roger. 'I was bored shitless, so I stupidly went out on operations with them that weren't anything to do with Australians, and I got caught up in a firefight. They were really good—I didn't have much time for the Americans elsewhere, but the Rangers were good soldiers. We were honoured by the Republic of Korea. I didn't even know the Koreans were there.'

In fact, more than 300,000 South Korean troops fought in Vietnam—compared to Australia's 60,000—and there were rarely fewer than 50,000 Koreans there at any given time. They had their own area of operations on the eastern coast of

A SAPPERS' WAR

South Vietnam. About 5000 South Koreans were killed and 11,000 were injured during the war. To even things up, North Korea also sent fighter planes to support the communist side.

Meanwhile, back in Phuoc Tuy, road-building wasn't the only civil engineering project with a military purpose—far from it. Land-clearing—cutting huge swathes through the thick bush and rainforest that offered concealment to the enemy—had a huge tactical impact. But like everything else in war, nothing is ever as simple as it sounds.

17. LAND-CLEARING

Just as road-building doesn't exactly leap to mind as a military priority, land-clearing seems to have more to do with farming than fighting. However, it's one of those Sapper tasks that is absolutely essential to modern warfare yet gets obscured by all the flash and bash of combat.

Sometimes it's as simple as creating a space around the perimeter of your base so that the enemy can't get too close before they are seen, or cutting back the vegetation at the sides of crucial roadways so that improvised explosive devices are harder to hide. One of the most effective ways to combat experienced and effective jungle fighters is to take them out of the jungle—or, more realistically, to take the jungle away from them. For instance, one of the first land-clearing projects by the Sappers of 17 Construction Squadron was undertaken at the request of the US forces manning an FSB in the north-west of Phuoc Tuy Province.

'The FSB was resupplied by road from Long Binh via the highway that ran from Saigon to Baria, the provincial capital,' says Barry Lennon, a captain with 17 Construction from April 1969 to April 1970; Barry's brother, Warren Lennon, was the first commander of 1 Field Squadron. 'That road ran close to the sea on the western side of the province, but the eastern side of the road had heavy vegetation beside it. The Vietcong's D445 Battalion operated in this area, and the land-clearing—which involved cutting back a fire zone about 200 yards from the road—was intended to make it a little harder for them to ambush convoys.'

The men of 17 Construction would use only a basic D8 bulldozer to do this, but land-clearing was a lot more difficult when it was on a much bigger scale; the techniques, too, could be a lot more sophisticated. A lot of land-clearing was done for much the same reason that the failed Barrier Minefield was built: to keep the Vietcong in the Long Hai mountains, away from villages in the area, in order to stop them operating freely against Allied troops. So the plan was to cut down large swathes of the jungle and bush around the Long Hais and elsewhere, which provided a very effective canopy under which the Vietcong could move relatively freely, and to turn them into a physical barrier that would not only slow them down but would also leave them exposed.

The idea was to rip 200-metre-wide trails through the jungle, leaving 'windrows'—long rows of stacked vegetation, like

LAND-CLEARING

hay left to dry before baling, only much, much bigger. These would cut across the Vietcong's normal pathways to and from their lairs and would allow Allied forces to move quickly, if they needed to, across their supply routes.

The large-scale land-clearing technique developed by 17 Construction was simple. A ship's cable of up to 500 feet in length would be attached to the towbars of two D8 dozers. The D8s would then punch through the jungle about 40 feet apart, dragging the cable between them. The cable or chain quickly flattened smaller trees and undergrowth. Occasionally, though, a large tree might get caught in the bight of the chain. When that happened, a third D8 fitted with a Rome plough would attack the tree, first cutting it through and then using the blade to lift the cable over the stump.

The Rome ploughs were very sharp dozer blades that sat a few inches off the ground and were curved, so as to cut a windrow of shrubbery and foliage after each pass. They were designed to follow the contours of the land, and would neatly slice off everything at ground level, while leaving root systems intact to avoid erosion.

'These blades also had what was known as a "stinger" which was a sharp protuberance welded to the furthermost point of the blade,' says Ventry Bowden, a Sapper in Plant Troop, 1 Field Squadron, from April 1969 to April 1970. 'The operator would aim the stinger at the edge of a tree that the blade couldn't handle and penetrate it, causing it to split.' Eventually he'd push the blade against the tree and it would topple and then the chain would be back in operation.

'Sappers called the cable "the chain", finding some perverse humour in being able to write home and say they had been "dragging the chain all day",' says Barry. 'The most difficult thing about getting the land-clearing team into full operation was finding an idle piece of ship's cable that was 500-feet long. But the port of Vung Tau was just down the road, a detachment of 17 Construction was in residence there, and Sappers are infamous as scroungers. Enough said.'

Peter Aylett, who arrived in Vietnam as a plant operator (but who would return for a second tour as a member of a MATT) was involved in the operation in support of the American FSB and then moved on to land-clearing around the Long Hais. 'It was to deny the enemy access, plus a bit more,' he says. 'There were eight Caterpillar D8 bulldozers, and the plan was to move forward in straight lines in 200-metre square grids, then two dozers would hook on to the chain and pull all the stuff down. The other six dozers would be windrowing the last lot that we had pulled down. The windrows used to start from the outer edges and push it in for about 50 metres, and then start at the centre of the trail and push it out.'

The chains and cables varied in length, depending on what the Sappers could acquire from the docks at Vung Tau. They dragged in a loop, rather than in a straight line, so too much length wasn't an issue. 'That worked quite okay,' says Peter. 'I finished up as inside chainman on the big ones around the Long Hais. When you pulled the chain—a big anchor chain 120 feet long, 60 pounds to a link—it was very difficult to walk through once the scrub was down.'

LAND-CLEARING

Peter would discover just how difficult, cursing his comrades' handiwork both when he was attached to MATTs later in the war and before that, when he was promoted to lance corporal. 'It wasn't a good move, for the simple fact that the bulldozers used to go through in rather heavy jungle,' he says. 'I was a lance corporal by then, and when you're a non-commissioned officer you're not on the machines. The power group sat on the APC, so I would head back there, wanting to sit on there too. But it was, "Oh, no, Lance Corporal, don't sit here—go down there and tell that bloke down there to stop doing this or stop doing that. So I had to trot all over this stuff right down to that fella. I'd approach him with great caution because some of the blokes were a bit edgy; if you came up too close, all they'd see is your reflection and suddenly you'd have a self-loading rifle pointing up your nose.

'So you'd take a wide berth, get his attention, wave and then you would head back to the command APC. Then it would be, "Peter, go up there and tell that fella..." Oh, for stuff's sake! You would spend the whole of your shift time tromping over this road in the jungle—that's why I wanted to hand the stripe back in. "Put me back on a bloody machine," I'd say. "I'm not doing this for the rest of my time." But they wouldn't accept it.'

Rome ploughs were used without the chains in land-clearing too. Ventry Bowden operated D8 dozers that had been fitted with the special bush-cutting blades. 'Our task, as plant operators, was to strip the earth of any shrubbery, bush, trees and anything else that could be used as cover,' Ventry says. 'This allowed the American gunships—helicopters armed with

50-calibre guns mounted on each side, similar in appearance to the old Gattling guns—to fly low over the cleared areas and shoot at anything that moved. I'm led to believe that these gunships could cover a football field in a matter of seconds, with very lethal effect.'

A Sapper walking a safe distance behind the machine—and only in the track marks left by the dozer, because of the ever-present danger of M16 mines—would use a compass to guide the operator in making the initial 'cut'. After that had been completed, another four bulldozers would then follow. Each would move one machine's width in, to allow that particular section to be cleared, then they would move on to the next clearing task, and so on.

Barry Lennon was one of the first to realise the danger that plant operators were in from anti-personnel mines. 'We in 17 Construction had learned the hard way that D8s—or, more specifically, the operators of the D8s—were more than a little vulnerable to Jumping Jack mines,' he says. 'The mine, you will recall, consisted of explosive with a propellant under it that, when triggered, launched the explosive about four to five feet into the air. Then a delayed detonation occurred, sending shrapnel man-high through 360 degrees.

'The device was equally susceptible to triggering by a careless foot or a careless D8 blade. Unfortunately, the jump was high enough, and the delay just long enough, to pose a real problem for the D8 operator. The mine could also be triggered by falling timber behind or to the side of a D8. So it became necessary to armour up the D8s with steel plate to protect

LAND-CLEARING

both the driver and the machine. These armoured dozers were later used to remove the Barrier Minefield. The armour made for even hotter and dustier conditions for plant operators, but I never heard one complaint.'

Allan Preston was on board a D8 when it went over a mine. 'It was August 1969 and we were clearing land around the northern side of the Long Hais,' he recalls. 'This thing went off under me—it lifted me out of the seat and I came down on the gear shift gate area. A bit dazed, all I heard and saw was an almighty roar, and I was engulfed by flame in a split-second. The dozer was on fire around the fuel tank, and right where my SLR was, along with four fully loaded magazines.

'Ron Roberts, who was on a dozer 40 metres in front of me, ran back—even though we were not supposed to leave our machines when an incident occurred. He leaned into my D8, grabbed me by the shoulders and tried to pull me out. With that, I gave out a yell as I suddenly had a severe pain in my groin area. It turned out that my left testicle had got caught in the gear change gate, so we opened my trousers, turned it around and got out of there.

'When we got to our sergeant, Ron Snow, on his APC, they couldn't remove my ear muffs because it was too painful, so they inserted a bayonet along the side of my head to let the air in to release them. The fire on the dozer went out and caused no damage, but it had five grouser plates—the teeth of the track—blown off it. The belly plate had smashed the engine sump and two con rods were bent.'

Sergeant Snow, who sent both Allan and the dozer back to

Nui Dat for running repairs, was one of six Sappers to receive the Military Medal in the Vietnam War (although not for this operation).

Ventry Bowden vividly remembers an incident that didn't have quite such a happy ending. 'One afternoon, after a day of clearing, I was pushing up a protective wall in a large circular formation with my dozer, when I noticed a small group of infantry soldiers walking through an area that I hadn't yet cleared with my machine. There was a loud *whoomp*, and time seemed to stand still. I will never forget that scene as the soldier—who, I came to know, was Rick Ashton—stood on the mine and was catapulted 15 feet into the air, amid the red glow of the explosive, dust and percussion, and crashed back to earth writhing in agony.

'The other two soldiers with him were blown out to the side of the blast, but Rick seemed to be the worst injured, with horrific damage done to his legs. This was another catastrophe that could have been avoided, if the Barrier Minefield had not been laid in the first place. The good news to come out of this unfortunate incident was that Rick Ashton survived the dreadful blast. I met up with him a lot of years later, and we still remain friends.'

The Long Green, an area to the east of Dat Do and southeast of Nui Dat, was where the Sappers first tested their clearing technique for tactical trails. It presented a more complex challenge for a number of reasons. 'The difficulty was a slight elevation in what we discovered very quickly was terrain akin to what Victorians would call "wet wallum",' says

LAND-CLEARING

Barry Lennon. 'The water table was close to the surface but not close enough to look like swamp. The surface looked firm but any equipment of any weight quickly broke through and became bogged in the underlying goo. However, the Green itself had sufficient elevation above the water table to take equipment and to support quite a thick growth of vegetation.'

This vegetation provided a covered approach from east to west for Vietcong resupply from the village of Dat Do. The land-clearing operation was intended to deny the Vietcong this covered approach, and to expose them to artillery fire from the Horseshoe. Initially, it was thought that a 200-metre-wide swathe cut from north to south, roughly through the middle of the Green, would do the job. The task was given to 17 Construction's land clearing team (LCT). But within minutes of unloading the first D8 on the shoulder of the road from Baria to Xuan Moc, the team knew they were in trouble.

'The D8 broke through the crust and became hopelessly bogged, which is hard to do with a crawler tractor,' says Barry. 'There was no option but to build a causeway across the wallum to ferry the machines to the higher area of the Green, where the clearing was to be done. From memory, I think we laid down a road about half a mile long. It was made of about eight inches of lateritic material from a nearby area that was quickly developed as a quarry. The road was sufficiently strong that it spread the load of the D8s as moved onto the firmer ground of the Green.

'Methods for filling some borrowed dumpers were a little rough. Without front-end loaders, but with a few idle dozers to spare, it did not take long to get enough fill to do the job. But I'm glad that there were none of today's health-and-safety inspectors around at the time.'

Initially, it was expected that there would be considerable reaction from the 33 NVA regiment that was understood to be to the east of the Green, as well as from other Vietcong units nearby, recalls Barry. So the members of the LCT, who were also able to call on artillery support from the Horseshoe, were supported by a troop of Centurions, a troop of M113 APC mounted cavalry, and a company of NZ infantry. An M113 command post was also provided for communications with the squadron headquarters at Nui Dat and with the Task Force, too, if needed. The Sapper team consisted of plant operators, a section of field engineers to work the timber and build the facilities of the FSB, and occasionally men from the Royal Australian Electrical and Mechanical Engineers, who were needed from time to time.

'A little Sappernuity allowed us to make reasonable use of the access road for resupply, without too much risk of it being mined by the Vietcong,' says Barry. 'Around the same time as this road was being built, both 1 Field and 17 Construction were challenged to come up with something to stop the Vietcong from mining known resupply routes around the province. Dirt roads were very susceptible to being mined. Our solution, in the end, was to spray a layer of Pena-prime liquid cold bitumen on the road. Then, in the morning, we would check it by helicopter flyover to see if any of the road had been interfered with. Pena-prime was such a messy material that any attempt

LAND-CLEARING

to bury a mine below the road surface was readily detected. There were ways around it, of course, but our local Vietcong did not seem interested or innovative enough to work these out.

'Perhaps the Vietcong did not really care about what we were doing, or perhaps they simply didn't want to be outside our FSB when someone pressed the button on a 20-pounder, but we never received any unfriendly fire in the four to five months I was living in the Long Green. In fact, we had very few incidents while we were there. The worst was probably what was said to have been a Vietcong ambush of my New Zealanders but which I suspect was really an ambush by a "friendly" ARVN battalion that was supposed to be working in a tactical area to the east, towards Xuan Moc, but which had got itself lost.

'Only once did we have to call in a "bushranger" airborne gun platform, and that was mainly to cover our cavalry, who liked nothing better than to be rushing off into the scrub in their M113s without remembering to load up some infantry first.'

Lieutenant George Hulse commanded the 1 Field Squadron LCT from January 1969 until the start of construction of the Nui Dat dam, in late February of that year. In January 1969, the LCT was located at the site of an old French fort, Phu My, five kilometres along Route 15, the main supply route that linked Saigon to Hoa Long and beyond. The area is adjacent to the infamous Hat Dich area and was well known for Vietcong and NVA activity. The fort was manned by a unit of the army of the Republic of Vietnam, under the command of an ARVN lieutenant. The LCT's task was to clear all vegetation along Route 15 for about eight kilometres in the area of the fort, to a distance of 300 metres each side of the road's centre line. The terrain was flat and the weather was fine, and good progress was made by the LCT. George Hulse still has vivid memories of the operation.

'At night, when the planties were trying to get some hard-earned sleep, the shed in which we were housed would become the focal point for countless numbers of large rats,' he says. 'They were after our food or anything that they might bite their way through to take from our packs. These rats would fight over a captured item—it was not unusual to switch on a torch and see four large rodents all gripping a piece of food, all squealing loudly as they tried to claim it for themselves. The four of them would jerk around in circles as they fought over a piece of food such as ration pack chocolate, which they loved. During the day, the rats would disappear into the earthen bunds built around the fort. We counteracted the rats by adopting very clean living standards—sweeping and ensuring that no scraps were left as an enticement. It worked after a while, and the rats left us alone.

'One morning, our infantry protection platoon was forming up to escort our D8 bulldozers into a new position to continue land-clearing when some plastic was seen protruding from the ground—and one young infantry soldier was standing on it. He was told to freeze, and the remainder of the platoon was cleared from the area, which was checked by prodding with bayonets. The lone digger who was standing on the mine was then rescued, and the mine and its firing mechanism and batteries were removed from the ground. The discovery was timely. The next vehicle to go over that section of track was a Land Rover carrying Corporal "Snow" Evans and driven by Sapper "Shorty" Yates. They escaped a twenty-kilogram bomb made with Chinese explosives.'

Shortly after that incident, the ARVN hunted down and killed a spiny anteater, which they took back to the fort. That night, Father George Widdison celebrated mass at the fort's little Catholic church. Some of the cooked anteater meat was offered to him, the ARVN officer and Lieutenant Hulse for the evening meal, as recognition of their good luck with the landmine incident. History does not record how the spiny anteater tasted.

However, this was war against a dangerous and resourceful enemy, and these amusing incidents were just circuit-breakers in a life in which caution and awareness could easily transmute into anxiety and fear. 'During the night of 9 February 1969,' George Hulse continues, 'the LCT occupied a part of the perimeter together with our cavalry and infantry protection units. In the early hours of 10 February, the engineer machine-gun post saw suspicious movement and we stood to. Four enemy soldiers were approaching the engineer position, and I requested permission from the infantry commander to open fire. Permission was granted. We waited until contact was unavoidable, then I ordered the post to open fire. One enemy soldier was killed instantly and the other three were very seriously wounded.'

One Sapper would be very badly affected by this incident, although it would take 40 years for the effects to be fully realised—we will come to his story in a later chapter. But it does illustrate that even though the Sappers of the LCTs and other construction units may not have been burrowing down tunnels, clearing bunkers, leading infantry through mine-infested jungle or riding on tanks and APCs, they were in every sense right on the front lines.

Later in 1969, a major land-clearing effort—Operation Beaver Dam VII—was launched in the foothills of the Long Hai mountains. Working through monsoons and over treacherous terrain, 1 Field Squadron (and later 17 Construction) cleared almost 6000 acres of jungle while under constant attack from rocket-propelled grenades and landmines. Sergeant Ron Snow won the Military Medal for his courage and leadership during the operation—one of only six Sappers to do so.

These land-clearing planties weren't simply dozer drivers—they were soldiers, with all the dangers that entailed. They were carving out a strategic advantage, and that made them a target for everything from snipers to RPGs and landmines. Inadvertently, these Sappers were also helping to build the economy of a future Vietnam, since the land that was cleared was often fertile enough for farmers to cultivate. As we'll see in the next chapter, one side-effect of war can be the building of infrastructure for peace.

18. BUILDING BRIDGES

There's nothing brings a gleam to the eye of an engineer—army or civilian—quite like the sight of a bridge. It can be as big as Sydney Harbour's or just a little footbridge that allows villagers to get to their fields or fresh water. The bridge is engineering in its greatest expression—defying nature, geology and geography to make the world an easier place in which to live. It's not hard to imagine that the very first feat of engineering was a log placed over a stream so that primitive man could get access to somewhere previously denied him.

Peter 'Pedro' Kelly, a member of the crew that built one of the biggest bridges in Vietnam at the time, recalls the massive logistical effort that went into preparing for its construction. He watched in amazement as a giant Sikorsky helicopter not only delivered the pre-prepared beams but actually dropped them into place 'like Meccano'.

'I had only been in country a matter of days,' Peter says, 'with barely enough time—or so it seemed—to unpack my "bags universal" and get myself issued with a weapon, before—with what I can only deem indecent haste—I was in the middle of a truck, in the middle of a convoy, in the middle of Phuoc Tuy, and in the middle of bloody nowhere. Apparently, we were off to build a bridge with 23 roots and sing a song about a chick called Rye in a place called Swan Mock. Well, like all other "reos"—or reinforcements—I didn't have a clue what was going on, and that included the lingo that was being thrown around, whether it was official military jargon or unofficial Sapper slang. By the end of the job, I was smart enough to realise that I had just built a bridge along Route 23, where it crossed a tributary of the Song Rai river on its way to the town of Xuyen Moc.'

Peter admits the whole operation was something of a mystery to him. Troop commander Captain Peter Knights had given the Sappers a briefing before they set off but many of them—Peter included—were too raw and readily distracted by the nervous excitement of heading off to a major operation to pay too much attention to the details. 'Less than a week earlier I had been at the bar of the Railway Hotel in Liverpool having a few farewell beers with some mates, and now, in what seemed like the blink of an eye, I was in a completely foreign land,' Peter recalls. 'Actually, it may as well have been a foreign planet. It was all pretty exciting, as well as being a bit scary.'

Peter and his mates arrived at the west bank of the river

and hoochied up on one side of Route 23. On the other side, the ARVN had set up an FSB to protect the new bridge and, before that, the men who would build it. To err on the side of safety, the Sappers were also protected by a troop of APCs from 3 Cavalry, and an infantry platoon.

'I was actually trained as a plantie, and by some bad throw of the dice I had been put into 3 Troop until a position in Plant Troop became available,' Peter recalls. 'It took about two months for that to happen. So it was with much interest that I watched the plantie on the TD15 that had gone out with us push up a perimeter bund to make an NDP around the hoochies.'

Peter was so engrossed in this that he failed to notice that all the old hands had already grabbed the best spots to hang their hoochies and he ended up bedding down right on the edge of the bund. He could only watch with envy the smart pragmatism of a Sapper from 21 Support Troop, who was in charge of a Patterson 6B water trailer. Experience told; within minutes he had slung a hammock from the trailer and set himself up a little kitchen.

'By the time he finished it looked more like a gypsy caravan than what they had tried to teach us to use in corps training at the School of Military Engineering,' Peter recalls with admiration.

The learning curve for everyone was steep, and watching the old hands establish their creature comforts was the smallest part of it. Peter vividly recalls the big day the Skycranes arrived with the main beams.

'Now, this was another whole new experience again,' he says. 'Before being called up, I had been a fully qualified crane operator working at the Whyalla docks in South Australia, so there was not much you could teach me about cranes. But I had never heard of a Skycrane. I thought everybody was having me on.'

In fact, these were giant Sikorsky helicopters, designed purely for lifting large and heavy objects by cables and will be familiar to modern Australians as the Elvis crane used to fight bushfires.

'One morning, there was this almighty noise over behind the jungle, and then this huge thing—like a praying mantis—came flying in. Now I understood the term Skycrane. Hanging beneath its skeleton frame was one of the large steel girders for the bridge, and from that, at either end, were two ropes, making four in all. Jerry Neale and I were ordered to grab the ropes on the east bank and guide the girders into place.'

The army was aiming for maximum publicity for this bridge and its construction, so they had arranged for the photographer from the army newspaper to be on hand to record events.

'We'd been warned that the downdraft from these choppers was in excess of 100 miles per hour and was quite dangerous; being just 20 feet below didn't help much. Just as the photographer was about to take the shot, Jerry got blown down the bank and just stopped short of the water. The whole team, including the loadmaster in the chopper, got a good laugh out of that. By lunchtime, all the girders were in place. I must admit, it all fitted together like a Meccano set. The fellas back at 21 Support Troop who had done the prefab had done an excellent job.'

The troop then got stuck into laying the decking, which was hard work, as all the timber—12-inch by six-inch planks— had to be carried out onto the girders by hand. On top of this, it was the wet season, which made things just that bit harder. The ground under the top crust was quite soft, and it wasn't too long before the TD15 bulldozer got bogged.

'The cavalry fellas thought this was great, especially as they were called upon to help pull it out,' says Peter. 'They did this while bragging about how great the APCs were because of less footprint pressure and so on. Within an hour, all three APCs were up to their belly plates, and so the cavalry guys had to come crawling to the planties to get them out. It was the only time I ever worked with so many Sappers at any one time—about 15, I think,' says Peter. 'However, that was typical of being in 1 Field Squadron: you never got to know a lot of people. Still, the thing that impressed me most was the bloke who knew how to be a gypsy right in the middle of a war.'

Peter Kelly's 'Meccano set' bridge was pretty impressive but it wasn't a patch on the monster that Robin 'Bob' Summerville, a lieutenant with 1 Field Squadron from July 1970 to July 1971, remembers. Bridge 6 was no ordinary bridge built to support military operations. It was the largest in a line of highway bridges, built by 17 Construction Squadron, on a row of concrete-bearing piles carrying 80-foot-long beams and reinforced concrete decking.

Route 2, which ran north from Baria to Phuoc Tuy province's border, was almost fully sealed by 17 Construction Squadron by about 1970. Route 23, running east from Baria to Xuyen Moc, was its poor cousin. It was sealed from Baria to Dat Do but became a goat track thereafter. And it had the geographical disadvantage of crossing all the major watercourses in the province, whereas Route 2 ran parallel to them. So bridge-building was vital if Route 23 was to be brought up to standard.

'As the number implies, Bridge 6 had five predecessors across minor streams,' says Bob. 'All of them were of reinforced concrete, ranging from oversized culverts to two-span steel and reinforced concrete structures. Where Route 23 crossed the Song Rai, however, the squadron had to build the largest piece of vertical construction attempted during that time in Vietnam.'

Peter's Kelly's bridge, innovative and challenging though it was, was a stroll in the park compared to Bridge 6. Designed in 1970 by 198 Works Section, Nui Dat, to replace a failing double-double Bailey bridge, it was a three-span, 40-metre bridge supported on two 11-pile bents. The 7-metre-wide concrete deck was supported on seven steel beams over each span. The centre span measured 27 metres and was carried on 45 by 80-centimetre rolled steel joists. The landing spans were supported by more modest 50-centimetre beams, each of which was 7 metres long. And all of this was to be built in a war zone, well beyond the protection of Task Force guns, and using only the standard equipment that a construction squadron possessed.

BUILDING BRIDGES

A works order and a set of seven plans arrived at 17 Construction Squadron headquarters in November 1970. The squadron's construction officer had worked with 198 Works Section to refine the design and to become familiar with the plan. He was to be the 'quality control' for the project. The job of building the bridge fell to 9 Troop, along with elements of Plant Troop and HQ Troop. In all, some 35 men with a range of trades—from field engineers to surveyors, from carpenters to plant operators, welders and painters—were involved in building the bridge.

The workforce was split roughly evenly between Nashos and regular soldiers, recalls Bob Summerville. They were housed in an NDP next to the bridge site. NDP Susan, as it was known, had a triangular bund of about 50 metres on each side, with a strong point on each corner. There was a tower that was manned during daylight hours and that afforded a view over several miles of the Song Rai floodplain. The NDP had showers, a kitchen and a blackout hut for recreation. Sleeping accommodation was halved Armco culvert sections built into the bund and covered with sandbags. Pairs of these formed a 'hutchie', each of which had its own sandbagged firing bay on the bund. While the workforce concentrated on the task in hand, the evening meal was delivered daily by helicopter.

A section of APCs from the cavalry squadron established standing patrols around the area during the day and boosted the strong points at night. An under-strength South Vietnamese Regional Force platoon lived at and guarded the old bridge. NDP Susan's defences were never tested, but it was a tough ask of the bridge-builders to spend long days on construction and longer nights providing their own security.

The area around the river crossing was a minefield that had been laid by the South Vietnamese Army. Before any construction work could begin, 1 Field Squadron's mine clearing team had to clear the site. Mini-teams picked up some 600 of the 1000 mines in the field, and the rest were buried. The work took a month. The 1 Field Squadron mine clearing team's progress is recorded laconically in the squadron's operations log. The job was 'completed without incident'.

The piers for the bridge comprised 11 steel pile bents, each topped with a concrete pile cap. Pile-driving began using the time-honoured drop hammer system, not hugely different from the technology used to build Watson's Pier at Gallipoli. The squadron obtained a diesel pile-driving hammer from a US Army piling company in Saigon. Its relentless beat drove piles into the silty floodplain with alarming speed—at least, until one pile hit an obstruction and broke. Extracting the obstacle and the broken pile, and then designing and building a workaround, took a little time and a lot of Sappernuity.

The main span beams arrived onsite in 10.3-metre lengths. A fabrication yard was established on the eastern approaches to the bridge. There, a beam was cut in half and the halves were welded onto each end of another beam. All the welding was done by hand, with some 8 metres of three-quarter-inch fillet weld on each beam. The welds were all X-rayed and found to be entirely without fault. The welders were mostly national servicemen plying their civilian trade, albeit under much tougher conditions.

Positioning the main beams was done from the original Bailey bridge. Each beam was placed on the back of two trucks—one facing left, the other right—and driven onto the old bridge. From there, Skycranes lifted it up and over the Bailey trusses and placed it on the new pads.

'It was an exciting and photogenic process,' Bob recalls.

The bridge was a major achievement, especially considering the logistics, teamwork and ingenuity it required. While 17 Construction Squadron had a batching plant and a fleet of transit mixers, the distance from Nui Dat and the state of the roads made the delivery of good-quality concrete difficult and dangerous. As a result, more than 140 cubic yards of concrete for the bridge was mixed by hand. Tests at Long Binh showed the quality of the work and the squadron works reports record that 6.5 cubic yards of concrete was being mixed and placed to a trowel finish in six hours by a crew of eight men.

Formwork and reinforcing steel was prefabricated at Nui Dat and delivered to site. Cement was delivered in two-tonne bags direct from Vung Tau, as were gravel and sand from Long Hai and Nui Dat. Fill for the NDP and the new road approaches was taken from a quarry established at Nui Nhon, about two miles west of the bridge site. Bridge 6 was such a remarkable achievement that it became a 'must-see' on the Indochina tourist route, particularly when the main beams were being placed.

'Everybody from the Minister for the Army to the Task Force

commander came to visit,' said Bob. 'Some offered advice; all took photographs. And some even stayed overnight.'

After five months, work on Bridge 6 was completed a few days ahead of schedule, in late August 1971. NDP Susan was disbanded and the squadron set about packing itself up ready to return to Australia. Not surprisingly, the project was forgotten in the excitement of going home. There were rumours that the bridge was attacked shortly after the departure of the Task Force from the province, but we know for sure that the deck was replaced by Vietnamese civilians in 1983. It is difficult to believe that, with the quality of the workmanship that went into the bridge, it would have failed within 12 years without a little help from explosives. In any event, the bridge was ultimately dismantled and replaced in 2006.

By that time, the economy of the (now) Baria-Vung Tau province had grown to such an extent that Bridge 6 constituted a bottleneck on the wide highway, now known as Route 55. The new bridge is slightly longer and twice as wide, and is made of reinforced concrete. It took Vietnamese contractors a year to build. All but one of the original piles was extracted, and the beams now see service around the province in bypass bridges.

Bridge 6 was a fine example of Sappers undertaking a huge project that a civilian contractor might consider a challenge—and doing so under constant threat of attack, using only the equipment, materials and manpower available in a war zone. But sometimes speed rather than size was the critical factor, as when, in the early hours of the morning of 2 August 1969,

Vietcong National Liberation Front forces blew up a bridge on Route 15, which crossed the Rach Hao river just south of Baria.

'The Rach Hao was part of the maze of waterways that were a feature of the landscape between Vung Tau and Baria,' writes Vince Neale in the *Phuoc Tuy News*. 'The first Alan Brown and Max Dadd knew of the event was when they were both dragged out of their slumber to drive officers to the site for a recce. Max was with 17th Const. Sqn at Nui Dat and Alan was with the detachment in Vung Tau, so Alan drove to the south bank and Max to the north bank.'

By 7.30 am Major John Wertheimer was meeting with a US colonel of engineers at the site, and it was decided to put in an American M4T6 floating bridge. This was a bridge that the Aussie Sappers had no experience of, so the US engineers sent along an officer and eight sergeants for technical advice and supervision, while 17 Construction Squadron would supply the labour. By 5 pm that afternoon the bridge had arrived by convoy from Long Binh. Waiting for it on the north bank were the Sappers from 9 Troop who were to do the manual work. In the meantime, tip-truck loads of road base had started to be piled on both banks for the approaches.

By 6.30 pm, just as the sun was going down, work began on the bridge. 'It was fortunate that all Australian sappers had basic skills in bridge building and these were easily transferable to the construction of an entirely unfamiliar portable bridge,' writes Vince. 'By the end of the task it was more than obvious that a mutual rapport had quickly developed between the US and Aussie Engineers to the extent that a bridge that would normally have taken two or three days to construct, was done in less than 24 hours.'

Work was accelerated by the use of an Austin-Western crane, which was a first even for the Yanks. It had arrived via a landing ship medium (LSM) just before midnight, when construction was already well underway. It was dangerous work. The equipment was unfamiliar to the Sappers, who had no tried and tested routines to fall back on. On top of all that, for most of the night it was pouring with rain.

The whole operation was a massive exercise in cooperation between different units and different armies. Electricians from the squadron provided lighting for the site, while D Company, 9 RAR, along with a troop from B Squadron, 1st Armoured Regiment and one from B Squadron, 3 Cavalry Regiment, provided protection.

'Sappers from 9 Troop worked for an exhausting 15 hours straight and by first light the next morning most of the work had been done,' writes Vince. This was no small bridge: in the water it was 60 metres long, and with its end spans attached the total length was 80 metres. No doubt the 9 Troop Sappers would have pushed on if they had been ordered to, but they were relieved at daybreak by another troop of Sappers, who finished the job.

By 3 pm, the bridge was completed; just 20 minutes later it was opened for traffic. In less than 24 hours—half of that in the dark, and in heavy rain—a rubber-raft floating bridge, 80 metres long, with diversion tracks and capable of carrying

BUILDING BRIDGES

60 tonnes, had been built. It was a triumph for 17 Construction Squadron.

But there was one more hidden danger to deal with.

'Before the bridge had been delivered a careful check had been made to ensure there were no mines or booby traps at the site, and none had been found,' writes Vince. 'Three days after the bridge had been declared open someone spotted a suspicious object in the water jammed up against the rubble. On the afternoon of the 7th August a combat engineer team (CET) from 3 Troop was ready-reacted to the site to investigate the object. It turned out that one of the original charges used to blow the bridge had not detonated and had slowly made its way to the surface over the next few days.'

The Sappers from the CET used a grappling hook made from four-inch nails to pull the package clear. Captain Peter Knight was with the CET and made the decision that the charge would need to be removed and dismantled, as to blow it *in situ* would more than likely destroy a large part of the new floating bridge, which sat only 100 feet away. The bomb consisted of approximately 250 pounds of Chicom plastic explosives, very neatly wrapped in black plastic sheeting and expertly bound by white nylon cord. Attached to this were a number of lozenge-shaped red plastic floats. The detonator had been placed inside a condom, which was placed within a primer made from a small round fish can filled with Chicom plastic explosive; it had then been inserted into the main charge. The detonator was disposed of immediately, while the rest of the charge was loaded into the back of a Land Rover and taken back to Nui Dat for disposal.

'It should not be underestimated as to how dangerous a situation this had all been,' writes Vince. 'The Sappers from the CET could easily have met their end as they struggled to clear the charge and dismantle it. So too could have all the Sappers from 17 Construction who worked on the bridge site while all along, lurking in the rubble more or less at their feet, was the highly unstable 250 lb charge of CHICOM. If it had gone off the explosion and flying debris would have been massive.'

The bridge at Rach Hao was a fine example of how effective and adaptable Sappers could be when the pressure was on. It also illustrated the immense range of Sapper tasks, from building a bridge to dismantling a massive bomb—that sometimes had to be done at the same time and in the same place. In Vietnam, it seemed there was nothing that Sappers couldn't or wouldn't do to get the job done.

19. HEARTS, MINDS... AND A LITTLE BEAR

There was a popular phrase bandied around Vietnam during the war: 'the battle for hearts and minds'. It must be remembered that the Allied forces in Vietnam were not an army of invasion and occupation; they were ostensibly there to help protect the South Vietnamese from communist North Vietnam. That they had come at the invitation of a profoundly corrupt government, which the United States had helped to install and which was routinely torturing and killing its own citizens, is another issue entirely. And the fact that a lot of Vietnamese civilians didn't want them there was fairly self-evident.

One of the biggest tactical issues that the Allies faced was that the Vietcong had infiltrated large areas of the South but could only operate with the approval and collusion of the local villagers. So, while there was a very serious armed conflict going on, there was also a program of 'civil affairs'—an effort to build infrastructure to improve the daily lives of rural Vietnamese in particular, in the hope that they would decide that supporting the Vietcong wasn't such a great idea after all.

Even the Vietcong could see the value of this work. They firmly believed that the country would soon be theirs to run, and a new road or bridge or school would be just as useful to them as to the Allies. One of the most telling examples of their acceptance of civil affairs is the legend of the Little Bear of 17 Construction Squadron, which basically had a free pass from the Vietcong and NVA.

Seventeen Construction was deployed to South Vietnam as part of the first Australian Task Force in 1966 and remained there until 1972. Major Peter R. Lofthouse first contacted the Behr-Manning corporation early in 1970 after he saw an ad for its transparent adhesive tape; the caption under the little bear logo read: 'A Little Behr Will Fix It'.

Peter was looking for a way of celebrating the squadron's 21st anniversary in September that year. 'The engineer corps had long had the nickname of "Ginger Beers",' he says. 'With a bit of a play on words, the slogan could be adapted to "A Little Ginger Beer Will Fix It". Behr-Manning were in, and once official approval was granted—the request going all the way to Canberra, apparently—a large package arrived containing transfers that had been especially produced for 17 Construction by the company.

'When I was visiting Canberra a couple of years later, I was advised by a senior officer of the intelligence corps that there was evidence to suggest that the NVA had issued a decree that anyone displaying the Little Bear was not to be targeted, as they were doing work to help the country and that this type of construction would be of great benefit when they eventually assumed control.' Peter adds that it's now clear that 17 Construction Squadron had been under some form of protection in Vietnam, as the unit had suddenly experienced a change in the attitude of the local populace. This was borne out by the documents captured at NDP Garth (see Chapter 16).

'Following the introduction of the new logo, there were only two incidents causing injuries to members of the unit,' he says. 'The first was when a driver, despite being repeatedly instructed to stay on the road, decided to drive off the main surface to turn his vehicle around. He ran over a large mine, which blew the cabin off the Mark 5 tipper and threw the driver up over the steering wheel, causing serious injuries including a broken pelvis, broken collarbone and pierced eardrums.

'The second incident involved an up-armoured D8 dozer that hit a buried 500-pound bomb during land-clearing operations. There was no damage to the operator other than a hell of a fright. The engine was blown upwards and sat across his lap, luckily being supported by side panels so that the full weight of the motor was not on his legs. Even so, he was trapped in his seat for some time until the engine could be raised to allow him to be freed.'

Significantly, in neither case was a 17 Construction vehicle specifically targeted—it was sheer bad luck. Conversely, another major incident occurred while the unit was upgrading a road north of the Bin Bah rubber plantation, at the base of the Long Hai hills. A large cart loaded with logs was overturned at the entrance to the quarry, preventing any of the tip trucks carrying decomposed granite from leaving.

'At the same time,' says Peter, 'a wagonload of straw overturned just south of the job site, which would have prevented any of our vehicles from returning to the quarry. A short time after these accidents an armed convoy leaving from the Nui Dat base to travel to the Bin Bah rubber plantation was ambushed by quite a large enemy force. It was apparent that this was not just coincidence—the "accidents" had been staged to prevent 17 Construction Squadron vehicles from accidentally finding themselves involved in the firefight.'

Other units' NDPs in the province were subjected to attack during the same period. During land-clearing operations, there were many instances of the local villagers warning 17 Construction operators about hidden ordnances, from small mines to 500-pound bombs.

'Despite the fact that the small teams were working in rather remote areas of the province, without any form of protection other than their own weapons, they were not subjected to attacks of any kind, and in fact the local villagers quite often offered food and drink as a form of thanks for the work they were doing,' says Peter. 'It was common practice for senior staff of the unit to drive around the province between projects in open Land Rovers, often with only their driver for protection. Despite the inherent dangers, they were

HEARTS, MINDS . . . AND A LITTLE BEAR

never subjected to attack. Other units, however, when moving through the same areas, were required to travel in armed convoys, often with armoured escorts.'

Peter says it was evident that the local powers that be, including the mayor of the Vung Tau city council, were fully aware of the Little Bear and the work 17 Construction was doing throughout the province. Many village chiefs provided personal protection for plant and equipment, which at times had to be left in village areas—this would not have been possible, or even contemplated, before the introduction of the Little Bear. It says much about the mindset of the NVA and the Vietcong that they assumed they would prevail and therefore benefit from the improved infrastructure. It probably has more to do with self-belief than foresight but there's no doubt that Vietnamese of all political hues appreciated that the Allies, when they eventually went home, would leave some aspects of the country in better shape than when they had found them. So while the Aussies hadn't won their hearts, the enemy's minds were made up—all this infrastructure would be theirs sooner or later so they weren't going to disrupt its creation.

'Perhaps the most cost-effective "hearts and minds" work we did—in that it was simple to do, relatively inexpensive and had high visibility with the locals—was road-sealing,' says Barry Lennon. 'The roads through most villages, while they had a good subgrade from years of use, were unsealed and, in the dry season, very dusty. In the wet, they were simply dangerous and got cut up by heavy traffic and metal-rimmed cartwheels. A road could be prepared for sealing—usually just grading—in a matter of days; it could be sealed with a spray bitumen in just a few more days.

'My favourite "hearts and minds" job was, however, when we created a reticulated village water supply at Hoi My to replace the daily morning drudge of getting it bucket by bucket from the local village well.' Seventeen Construction had a drill rig, and finding the right place to drill was not exactly rocket science. Villages tended to be built where a local hand-dug well would find water, so drilling a well nearby rarely failed to find a source. 'But the real innovation,' says Barry, 'was the introduction of technology exploited in Australia for decades—the windmill—to lift the water to a storage tank. When we installed that, plus a tank on a tower, instantly there was a clean, gravity-fed village water supply. The locals thought this was just great.'

As Barry discovered, there were little jobs that meant a lot to villagers but could be easily accomplished if you had the right equipment. 'There was a huge boulder that blocked access to a village along the only viable road to it,' he says. 'The local way to remove it called for building a series of fires over it until, after several days or weeks, the constant heat expansion would crack it into more removable pieces. But a hole drilled down to the centre using 17 Construction's trailer compressor, and a CE primer positioned and tamped down the hole, was enough to do the job—and without damaging any surrounding buildings. Just a short distance away was a building that had to be sandbagged to ensure it survived the blast. As I recall, the sandbagging probably was not necessary. One of 17 Construction's powder monkeys had an uncanny ability to judge where to position the primer so that just a small blast would exploit fault-lines of weakness in the rock. The rock just fell apart when the primer was detonated. The kids were most impressed!'

Long Son island, just off the coast to the west of Baria, was believed to be strongly sympathetic to the Vietcong, and although the provincial army was present, no one was really ever sure if some of them, at least, did not favour the enemy. 'As part of the civil affairs mission of the Task Force,' Barry recalls, 'it was decided to build a medical centre in the village of Long Son on the eastern side of the island. Provincial government would man the centre on a "boat in, boat out" basis, as it was only a short trip from the highway linking Baria and Vung Tau. But there were also occasional visits from Task Force medical staff from Nui Dat and Vung Tau.'

Seventeen Construction was given the job of building the centre in the village. Although it wasn't a big job, the greatest challenge was the logistics of getting building materials to the island, which meant getting the timing right as, at low tide, there was no choice but to wait—or wade through several feet of mud. 'The construction of the centre went to plan and seemed to be appreciated by the locals,' says Barry. 'The head man threw a celebratory dinner to thank us when it was done. It was a pleasant experience all round—except perhaps for the later discovery that the main course at the dinner had been dog, which was not a dish with which many Australians were familiar... or comfortable.'

20. THEY ALSO SERVE...

So far we have met the Sappers who went down tunnels, deloused booby traps, laid mines and cleared them. We've met bridge-builders, road-builders and land-clearers as well as the guys who built medical centres and dug wells. Could there be any other tricks of the Sapper trade? Electricity supply seems like an obvious extension of their skill set but, amazingly, Sappers were also in charge of the landing ships used to get vehicles and tanks from sea to shore.

Remember that Sapper motto: 'Everywhere'. Sometimes, 'all over the place' would have been more appropriate, as in the disastrous attempt to land troops on a remote beach where secrecy was at a premium and planning took a back seat. It all started on the morning of 29 December 1967, when a platoon of infantry, along with men from 1 Field Squadron

THEY ALSO SERVE...

and 104 Signals Squadron, boarded the *Harry Chauvel*, a ship anchored off Vung Tau. The *Harry Chauvel* was a landing ship medium (LSM) and was designed to transport five medium-sized tanks.

As Captain Fred Hartnack, Warrant Officer (Class Two) Frank Johnston and Sapper Dave Clark told the *Phuoc Tuy News*, a total six officers and 46 other ranks made up the group on board for the mission. Five assault boats, along with their 40-horsepower Johnson outboard motors, had been loaded the previous day, to be crewed by Sappers from the *Harry Chauvel*. Captain Hartnack, the ship's commanding officer, had been instructed to investigate the possibilities of using a small ship for the purpose of 'operational logistic support'.

'He wasn't told what the Infantry task was, but it was assumed by the fact that it was only a platoon-strength unit led by a major, that it was reconnaissance,' writes Vince Neale. The *Harry Chauvel* sailed 43 nautical miles up the coast from Vung Tau to the south-east corner of Phuoc Tuy province, where the border meets the coast. 'Prior to the task nobody from the *Harry Chauvel* had been trained for an amphibious landing other than the normal loading and discharging of vehicles and cargo,' writes Vince. 'Certainly the sappers had never had experience or training in this sort of activity.' Although they were familiar with the operation of small craft, this too was limited to the basic training all Sappers received at Camp Sapper on the Georges River in Sydney, once they had left the School of Military Engineering.

As they neared the shore, they spotted a sandbar about

400 metres from the beach and Captain Hartnack decided this was too dangerous to negotiate so he anchored the ship in deeper water. The infantry colonel commanding the landing force ordered the operation to continue. The bow doors were opened, the ramp was lowered and the five assault boats loaded with all the gear were launched into the water. Life jackets were provided but were not worn by the troops, as they would have been an encumbrance.

'That's when it was noticed that the infantry were boarding with full combat kit, weapons, bandoliers, grenades, etc. as well as bum packs, and were wearing GP boots,' writes Vince. 'One of the lieutenants was taken aside and had it pointed out to him the danger of wearing all this gear. To his credit, the young officer consulted with the colonel and, fortunately, permission was given for them to strip their gear down to their fatigues although some of the troops kept the M60 belts and bandoliers on.'

The five boats then headed towards the shore. About 30 metres from the ship, they encountered the highest part of the sandbar. Sandbars are well known to mariners as being very tricky and dangerous obstacles, seemingly with a life of their own. When a swell approaches a sandbar from the seaward side, it sucks the water back from the sandbar, causing a trough. The swell then hits the bar as a wave. This is what causes trouble for boats, as it comes upon the stern and forces the boat into a sidewise movement. Quite often, it causes a boat to capsize.

'This is exactly what happened to all five boats at the same time, sending them all into a barrel roll,' writes Vince. 'Everybody was thrown out of all five boats. All equipment—weapons, radios, webbing, boots, and outboard motors—went straight to the bottom of the sea, which at this stage was between six and eight feet down. As well as the equipment, those troops wearing the bandoliers and belts also went to the bottom.'

Realising that the men in the water had little or no chance of getting to shore, not to mention those who had sunk to the seabed, Fred and Frank launched two inflatable rafts from the ship and, within a couple of minutes, reached the scene of the disaster. The life rafts were circular rubber rafts meant for protection rather than propulsion, with a tent-like structure on top and two hatches on opposite sides; powered simply by paddles, they were almost impossible to control. But there was a fairly stiff breeze, and it did all the powering.

Meanwhile, some troops had managed to cling onto the sides of the upturned boats; others had grabbed hold of the few lifejackets that had made their way to the surface. Some men had started to swim to shore but others were drowning. Frank dragged five men on board his raft, while another ten or so clung to the lifeline. The raft was then caught up in a rip and they were carried south for about 300 metres. Finally, they were dumped onto the shore—all safe.

Fred drifted over to the troops who were on the bottom, desperately struggling to free themselves from their weighty gear. This they managed to do, and they made their way to the surface. Then, loaded up like Frank's, Fred's raft was blown to the shore. Miraculously, the only injury suffered during all this was one broken leg.

'While all this was happening,' writes Vince, 'the *Harry Chauvel*, under the command of Executive Officer Bill Tindale, was on full alert, with all armaments on standby. Orders were given not to fire any weapons for the risk of injuring or killing any of their own troops. The only weapons to reach shore were one Browning pistol and one SLR that a dedicated digger had somehow managed to hang on to. There they stood, some 50 personnel, wet as shags on a rock with nothing but the greens they stood in, not even boots. They were unarmed in enemy territory and without doubt the enemy were not far away. Apparently it was attempted to get a line ashore to the ship by the ship's cook but to no avail. He was hauled aboard before the possibility of a further disaster.'

The colonel must have sent out some frantic radio messages, because about 90 minutes later a Chinook and five Huey helicopters arrived on the scene. One Huey picked up Fred and Frank and winched them back on board the *Harry Chauvel*. All troops, assault boats, life rafts and equipment were lifted off and flown directly to Vung Tau. The *Harry Chauvel* sailed to Vung Tau, arriving at 10 pm, when the remainder of the infantry and other units disembarked and the ship's crew was reunited. Nobody lost their lives, thanks in large part to the quick thinking of Fred and Frank Johnston—plus a good stiff breeze and an awful lot of luck.

When you learn that Sappers were in charge of small ships, it's probably no surprise to learn that they were also in charge of troop movements in wartime—although why this would be a Sapper task rather than, for instance, logistical support or intelligence is mystery. As Corporal Ralph Todd of 11 Movement Control Group, RAE, discovered, a simple request to make sure everybody got their fair share of R&R saw him caught in some of the most ferocious fighting of Australia's involvement in Vietnam.

'Much effort had gone into setting up a good R&R system with the Yanks,' says Ralph. 'Their Military Assistance Command Vietnam (MACV) allocated a certain amount of seats on their flights for Australian troops; however, it was crucial that the seats were filled. The Yanks had warned that if the allocation of seats was not used up then the entitlement would be withdrawn. As well as that, there was the fact that the R&R system was not working too well—the main problem being getting diggers out of the bush at the right time.'

Trying to match a combat schedule with the availability of seats on a plane meant that some were missing out completely, while others were being drawn out of the bush too early, ending up sitting around, filling in time, all of which was affecting their morale. Things got so bad that 1 ATF headquarters ordered that 11 Movement Control Group place an NCO at all the FSBs to ensure the smooth running of the R&R service. And that's how Ralph found himself on the very first convoy heading into the thick of one of the most ferocious battles of the war.

'FSB Coral was about a six-hour road trip from Nui Dat,' he recalls. 'The trip was undertaken over two days, with the first day being the long haul to the US base at Long Binh, where we bivvied up for the night at Bearcat, a large US military base east of Saigon, which was home for many military units, including airborne, marines, military police, communications, support and maintenance. The next day saw us up bright and early and practically on the heels of the men of 1 RAR, who had gone in to clear the area that morning. Unfortunately for them, they stumbled upon large forces of NVA who just happened to have bivvied up in that area as well. As a result, a fierce battle ensued.'

Ralph's job had three major components: handling the medivac, coordinating the transport (especially air movement) in and out of the FSB, and ensuring that the diggers who were to go on R&R were airlifted back to Nui Dat. 'Right from the start, the site was under constant attack from mortar and RPG fire,' he says. 'The pace and atmosphere was frenetic, to say the least. I was attached to 1 ATF headquarters and received my orders straight from there. Once the command bunker was built, I attended at least three briefings a day surrounded by all the brass. I was the only Ginger Beer there; there were others at the FSB but I never saw or met any of them. So it was up to me to look after myself.

'I was told to dig in, and this I did willingly. However, I was so busy with my tasks that after three days It had still only managed to scrape down 18 inches. All up, it took me two weeks to get down to the required four-feet depth. It wasn't that I was slack or anything, it was just that there was no time. To ease the burden on me, the units looked after their own medivac and I concentrated on directing the chopper traffic and seeing to the R&R boys.

THEY ALSO SERVE ...

'Most of the air movement was coming from the Chinooks bringing in artillery and small-arms ammunition, plus food and water. My call sign was "Grasshopper"—and a pretty apt one it turned out to be. FSB Coral would have been at least five or six times the size of the MCG and there were a number of landing zones (LZs) for the choppers to hit. I had to get myself to every one of them to guide them in with smoke bombs and so on. The Royal Australian Army Service Corps (RAASC) provided me with a Land Rover for the job, and it got a thrashing as I hopped from LZ to LZ. Most of the time the Chinooks didn't touch down; they hovered about 20 feet up and just dropped their loads. The RAASC boys then ducked in and quickly loaded up Land Rovers and Mark 3s and got the stuff distributed to the various units.'

Morale tends to be forgotten when, 40 years later, we look at the contributions that soldiers made to the war effort. Work such as Ralph's doesn't quite match the heroics of an individual acting above and beyond the call of duty under fire, but when you add up the cumulative effect of making conditions more tolerable for all the front-line troops, it's clear the least heroic job can have a profound impact.

Les Hillberg was destined to be an unsung hero from the moment he was called up for national service in 1969. He had already completed his apprenticeship as an electrician, so he applied for allocation to the RAE and was sent to Vietnam as a reinforcement electrician to 21 Engineer Support Troop (EST). Les admits to still carrying a little guilt from having served his time in the relative safety and comfort of the generator hut, while his mates were risking life and limb in very inhospitable conditions. But no soldier who came back to Nui Dat from patrol, looking for home comforts like electric lights and cold beer, would have begrudged Les his 'cushy' number. Even if a little reluctantly, Les concedes that 'I suppose I did do some good'.

Les was allocated to the power station in the infantry battalion's lines in the northern area of Nui Dat, and spent virtually his whole 12 months there. He lived there with the men of 7 RAR for the last half of their 1970 tour and with those of 3 RAR for the first half of theirs in 1971. The station required four operators: three electrician Sappers and one infantryman. The latter, who usually had no mechanical or electrical knowledge whatsoever, had to be trained up by the others. Any repairs or significant maintenance, of course, was carried out by the Sappers.

Being in charge of the power supply gave the electricians a degree of influence and status well above their ranks—something they rarely failed to exploit. For instance, Les and the other electricians always drank with the 7 RAR boys in the admin wet canteen. However, in November 1970, a new canteen was built; on opening night, Les was told by the barman that he and his mates were no longer welcome there. It seemed they wanted to keep the new boozer exclusively to themselves, apparently even barring members of other companies of their own battalion who had their own boozers.

'I suggested to the bartender at the time that I could wander up to the power station and turn off their electricity if they continued with that stance,' says Les. 'To which the barman replied that I wouldn't be game.' So Les walked up to the power station and switched off the feeder that supplied the canteen for one minute, then he switched it back on again. 'I walked back to the canteen and up to the bar, and the same bartender said, "How many beers would you like?" Everyone was welcome thereafter.'

Because the power station was run by four operators, who could provide rolling shifts of eight hours on and 24 hours off, they were pretty well tied down to the station and their tents. They were only allowed part-day excursions around the base, or the occasional 'shotgun' trip in the passenger seat of a truck going to one of the FSBs. Work at the power plant was tedious. Apart from taking hourly engine and load readings and refuelling the generator sets, plus the occasional oil change, the only real tasks were to shut one set down at about 11 pm each evening, leaving just one generator running overnight, and then to put another one back on before 5 am, when the cooks in the various messes began preparing breakfast.

About halfway through Les's tour, he and his companions decided to make their shifts 24 hours on and then three days off. This allowed the blokes to widen their horizons on their days off, and Les frequently jumped on a troop truck going to the Badcoe Club in Vung Tau or hitched a ride on a Huey heading to 'Vungers' from Nui Dat's Kanga helicopter pad. Sometimes he even hopped on a Caribou and flew up to Saigon for a few days.

'Of course, this activity was never mentioned to others,'

explains Les, 'as it wouldn't have been condoned. There were no records of absences, or passenger lists and so on, so we were effectively AWOL on these excursions.' None of them were ever caught, however, and it made the rest of Les's tour a great adventure.

Not being content with the 'two cans, per man, per day, perhaps' beer ration that the infantry had to put up with, Les cut a trapdoor in the timber flooring of his tent, dug out the dirt beneath, inserted an army trunk and lined it with foam sheeting, to make a very effective icebox. The trapdoor was concealed by a rug. 'Having befriended members of 9 Squadron RAAF in Vung Tau,' he says, 'I would hop on one of their Hueys, fly to their base at Vung Tau for the night, avail myself of their hospitality, then return to Nui Dat with the Huey contingent the next morning. I'd have a carton of beer under each arm, purchased from their canteen—all very unofficial, of course.'

It was easy to get a lift back to his tent from a passing Land Rover, especially when he offered the driver a few cans of beer. 'I got a block of ice every morning from the driver of the ice truck for a six-pack, so I always had cold beer available at any time of the day.'

Having secured an illicit supply of ice and beer, Les started up a 'crown and anchor' game—a gambling dice game. The timing was somewhat cynical, he admits, as he scheduled it for the grunts' pay night. The word was put around that any of the lower ranks could drink as much free beer as they liked—which cost Les ten cents per can—as long as they played crown and anchor for money. As Les provided the beer, he was always the banker.

It was a chargeable offence to gamble in the army, so Les used strips of Paludrine anti-malarial tablets as chips, safe in the knowledge that the grunts could not acquire them from elsewhere; each man was handed his one tablet every day at morning parade, and made to swallow it on the spot. Tablets for the powerhouse operators were given to Les in advance, but none of them bothered to take them so he had quite a stash. Money only changed hands when someone purchased or cashed in his chips.

'Occasionally, the duty officer would walk in unannounced, but there was no money to be seen,' recalls Les. 'Some of the diggers looked a bit under the weather, for only having drunk their permitted two cans from the boozer, but I always made sure each man only had one can at a time, and that all the empties were well hidden. No doubt the officers all knew what was going on but they could not prove anything. I made a bit of extra money out of that venture.'

When 3 RAR took over from 7 RAR, there were, as always, a few junior officers, posted to their first unit, who liked to throw their weight around. A few tried to get Les to put extra lighting and power points in their very Spartan tents, to which his standard initial reply was to quote the army regulations: 'One 40-watt bulb and one power point per tent.'

'Of course, this was a bit hard to do with a straight face inside the power station tent, which had fluorescent lights and a multi-outlet power board and various electrical appliances on the table, and usually every man had a stereo system of some description in his own corner,' Les recalls. 'A couple of these officers even said, "Corporal, I *order* you to install extra lighting and power points in my tent!" but that only made things worse, and I resolutely stood behind the regulations.'

A few reported him to his commander, but they got short shrift from a practical engineer officer, who told them that if there were complaints about the electricity supply, he would listen—otherwise, bugger off!

Eventually, the quartermaster of 3 RAR, who was a middle-aged man, probably a warrant officer, asked the same question of Les, indicating that he had a small refrigerator, electric jug and frypan he liked to use to cook his own meals, and that the 40-watt bulb was very dim. Upon receiving Les's standard reply about the regulations, the quartermaster looked around the interior of the power station tent, paused and thought for a moment, then said, 'Corporal, is there anything I can do for you?'

Les explained that he got around a lot and while it was compulsory to have your issued firearm with you at all times, his self-loading rifle was a bit cumbersome. 'The quartermaster asked if I would like to swap my SLR for a service pistol and holster, which he could arrange,' recalls Les. In return he got a fluorescent light and extra power points.'

All's fair in love and war, and schemes and scams are part of army life all over the world. But one incident of which Les is not particularly proud came when he missed his American flight home for R&R. Les had decided to return home for his

THEY ALSO SERVE...

girlfriend's birthday in April 1971, and he had booked his seat well in advance. On departure day, several soldiers were picked up from Nui Dat in a Caribou and taken to Saigon, where they all dumped their army bags in a pile outside a building at Tan Son Nuit airport. They were briefed by a clerk, who said they could go and have some drinks as there were a few hours to kill before the Pan Am flight to Sydney.

Les and another soldier did a bit of a bar crawl, and when they got back to the rendezvous point theirs were the only two bags there. The clerk stuck his head around the corner and said, 'You're in trouble now.' Les hadn't heard the clerk say that they were to be back by 1800 hours (6 pm), and had relied on his friend's assertion that he was certain the clerk had said 8 pm.

They were billeted at the Free World Centre in Saigon for two nights, and were lucky enough to be allocated two spare seats on a flight two days later. Of course, not arriving home at the expected time caused some consternation with the family. Les's father had given the army a very hard time, wanting to know where his son was. Ultimately, everything worked out okay. Les expected to be disciplined over the incident but nothing happened. Even if it had, you can bet he would somehow have wriggled out of it.

21. SAPPERNUITY

There is no dictionary definition of 'Sappernuity', so here's one that may serve: Sappernuity is knowing what needs to be done, and doing it, often under fire. Sappernuity is practical and pragmatic, yet imaginative and inspired, ingenious in the face of adversity and ingenuous in the glare of authority. Sappernuity is what happens when the impossible meets the indomitable and the unfeasible encounters the imperturbable. It's a blend of psychology, kidology and technology. It's the cold courage of men who know better than most the extreme dangers that they face, because they make *and* they break. Sappernuity is all of this, wrapped up in the scruffiest fatigues, a weary smile, a nod and a wink.

If you are looking for the embodiment of Sappernuity, you need go no further than the legendary Harry Buckley. 'Harry was an expert in the field of mines and booby traps,' says fellow Team member Peter Conway. 'We visited many village and hamlet outposts, and we accompanied soldiers of both the Regional Force and the Popular Force from these villages on night ambushes on numerous occasions. We had a couple of interesting days and nights at an outpost manned by the US Navy Seals on the coast. Harry was a great role model and had the ability, in a very short period of time, to earn the respect and friendship of his fellow countrymen, the Americans and the South Vietnamese.'

Another Sapper Team member, John 'Speedie' Sahariv, also recalls Harry well. 'Harry was well known for his bartering skills with the local South Vietnamese population,' he says, 'especially with items such as cigarettes, whisky and large fleecy towels, which we were able to procure in abundance via Dong Tam army stores. He used to disappear for a couple of days at a time into My Tho—a large city six to eight clicks east of Dong Tam—and always returned with a smile on his face.'

Harry may have had a larrikin streak, but he was a professional soldier through and through. 'He was hard-working, decisive and brooked no nonsense,' says Brigadier Frank Cross. 'Except when he declared comic relief time, which was usually to relieve stress for students on courses such as the warrant officers' qualifying course. It was then that his sense of humour contributed significantly.'

Very few Australians who served in South Vietnam can display citations from both the US and South Vietnam—Harry could. His legacy as a Sapper, a leader and a trainer still has an impact on the spirit and confidence of the men of the RAE.

Lieutenant Roger Cooke's memories of Sappernuity combine two key elements of the 'gift': an eye for a bargain and lateral thinking. During the 'tethered goat' operation at FSB Garth, the conveyor belt on the Eagle rock-crusher wore out; as it was not an Australian standard-issue issue belt, it was difficult to replace. However, a private American company, Rathmusson, Merchason and Knox (RMK), was constructing a highway through Baria. They had an enormous quarry and hotmix batching plant; most importantly, they had the conveyor belts that the Sappers required.

'These belts cost tens of thousands of dollars,' Roger recalls, 'but all the bloke at RMK wanted was four slabs of VB. The deal was passed on to the officer commanding the RAEME detachment at 17 Construction in Nui Dat, Captain Phil Kersley, who put the slabs in his Land Rover and set off for Baria. He got no further than the front gate of the Nui Dat compound, where the military police stopped him. Accusing him of dealing on the black market, they wouldn't let him through.

'But his tenacity and Sappernuity have to be admired. He hid the slabs in a truck that had hit a mine at Garth, put it onto a low-loader and drove it to the RMK quarry, where the deal was done and the belt was handed over.'

Trevor Shelley knows what happens when Sappernuity is used for evil rather than good, and there was one incident of which he is not particularly proud. 'This is known throughout the corps, but I can't say that I have gained any pleasure from that fact. But I would be in denial if I did not give this at least a passing airing. It's really amazing what one can do with a coconut tree. Over the years, I have seen it turned into soap and body oil, and the wood even turns pretty well on a lathe.

'But by far and away the most important thing for us—and it was a time when there was no beer, except the odd pallet of Swan Lager that had been left sitting on the beach at Vung Tau for too long—was that the coconuts can produce wonderful moonshine if you put in certain additives. We cut the tops off them, poured out some of the milk and put in things like raisins, sugar and God knows what else. Then we buried them for a couple of weeks.

'Out of the ten we buried, we only found six. The result was impossible to drink straight, but with some 7 Up and some added sugar it became quite moreish. The stuff really was rocket fuel, and in no time at all the job of getting boozed out of your mind was complete. It must have been 100 per cent proof. Of course, I don't have very clear memories of what came next.'

In Trevor's case, it was an incident that could have ended up in a court martial. He found himself standing beside the rather new troop commander's tent, pulling the pin out of a tear-gas grenade and lobbing it in.

'The shrinks say that it does not matter how drunk a man becomes,' Trevor recalls, 'he still will not do things that are against his principles. Thank God for that, because it could just as easily have been a M26 fragmentation grenade—we had plenty of those. The act done, it was straight to bed for me. In no time at all there was a midnight muster parade. I suppose it was a bit obvious that I was the only one wearing greens and had my boots on when we fell in. The senior NCOs made threats of denial of privileges if the perpetrator did not step forward, so what could I do? I bloody well stepped forward. A lot of the blokes in the troop later said that I should have kept my mouth shut and that they would have taken the group punishment standing on their heads. No doubt they were right, but the deed was done and I was under close arrest.'

The next day, still reeling from the moonshine, Trevor was marched into Major Florence's office. Because of 'bad behaviour', he was given 28 days in the nick rather than in field duties. This was spent in the two cells at the entrance to the Australian Logistical Support Group (ALSG) base in Vung Tau. 'I think I got the message after that, as I was released to normal duties in 26 days—two off for good behaviour,' says Trevor, who adds, 'It's a relief getting it off my chest, but I still remain somewhat disgusted with myself for the act of stupidity that the moonshine fuelled.'

While the message from Brian Florence got through to Trevor loud and clear, a recurring theme of Sappernuity seems to involve putting senior officers in their place. John Tick was no exception when a bumptious infantry officer decided to show him who was boss. It was the last quarter of 1971, about the time the Australian government announced the withdrawal of Australian combat troops from South Vietnam. John was asked to supervise the clearance of a section of the defensive minefield around the base at Nui Dat. A couple of armoured

D8 bulldozers were allocated for the clearance but the normal supervisors were not available. 'They were probably enjoying some R&R in Vung Tau,' says John.

'Basically, my task was to do a quick recce of the section of minefield to be cleared, which was about 100 metres by 400 metres, and then meet up with the mine-clearing team the next morning. I may also have allocated a couple of two-man Sapper teams to check the cleared area with mine-detectors.'

Captain Dennis Pegg, John's troop commander, warned him that those at Task Force headquarters were interested in the clearance; someone would probably drop in during the day to see how things were going.

'The next day, after briefing the team, I was alerted by one of the Sappers that a Land Rover was approaching, so we all turned and waited to see who it could be,' recalls John. 'We expected it to be the rep from Task Force headquarters. The Rover came to a stop, and out stepped a flunky captain—it was the Task Force's chief of staff, an infantry lieutenant colonel who was known not to be a lover of Sappers of any description, rank or race. He strode over to our group and asked, "Who's in charge here?"

'Hesitantly, I announced, "Me, sir." The colonel fixed me with a steely, withering gaze designed to subdue anything from a tiger to the Wild White Man from Badu, then said, "Are these people Sappers?" I replied in the affirmative. "I could tell," he said, giving a disparaging glance at the plant operators. There were dozer drivers and field engineers standing around, variously dressed in greens with the shirtsleeves torn off and the occasional trouser leg cut off at the knee; thankfully, all were armed with M16 or 7.62-mm SLR rifles and the odd M79 grenade-launcher.

'I gave the colonel a short brief of the task and asked him if he had any questions. "A few," he said. "Firstly, do these people always dress like this?" He was clearly expecting the Sappers to be clad in clean, starched, knife-edge pressed greens as if on an infantry parade. "Secondly, when will you finish the clearance?" he then asked.'

Knowing full well that they would finish about 3 pm—in time for the diggers to get back to their hutchies for a shower and then an early beer or two at the boozer—John advised the colonel that they would finish just before last light, by about 5.30 pm.

'Not good enough,' announced the colonel. 'I want it finished by 1630 hours at the latest, and I will be here to check so you had better be here too.'

'"Bugger!" I thought,' says John. 'Trust a grunt to interfere with a well-planned and ordered Sapper activity—and especially with the chance for a few cold ones later. I suggested to the colonel that we should pull back a couple of hundred metres behind the wire while the dozers got started clearing the top six inches of soil and digging slots to bury the mines and anti-lift grenades. I, for one, did not want to be ventilated by some high-speed hot shrapnel from an exploding mine or grenade.

'The colonel, having lost interest by this stage, jumped into his staff Land Rover with his flunky and drove off, no doubt

SAPPERNUITY

for morning tea at the Task Force mess, while we stayed all day clearing the minefield. We finished the task around 2.30 or 3 pm. After a walkover check and a sweep with the minedetectors, we left the area, which now had new and repaired fencing at front and rear and an access slot. The Sappers all went for a shower and a beer, while I had to return at 4.30 pm to meet the chief of staff.'

At about 4.15 pm John drove back to the minefield. He entered through the access slot and waited in the middle of the minefield for the chief of staff, as ordered. At exactly 4.30 pm, the colonel and his flunky arrived and parked near John's Land Rover.

'The colonel got out and strode over with his parade-ground-perfected stride. "Well, finished, have we?" he said, introducing an element of ownership into the exercise, as if he and his flunky had assisted in some way. "Yes, sir!" I replied as casually as I could without displaying dumb insolence. "OK so what is the probability of clearance of this minefield," he asks. Thinking "here's a go!" I replied, "About 70 per cent clearance. We can't guarantee any better unless we neutralise each and every mine and grenade."

'The colonel's eyes widened and his pupils dilated as he focused on my face. "And just where are we, in relation to the minefield?" he asked. "Smack in the middle," I replied. Let's see what he makes of that, I thought. The colonel nodded, did an about turn and traced his own footsteps carefully back to the Land Rover. He and his flunky drove off equally carefully, as close as possible to their previous wheel tracks. Clearly, they were not 100 per cent impressed with the Sappers' efforts.'

On arrival back at squadron headquarters to brief Dennis Pegg, John discovered that he had already had words with the colonel. 'He had a silly grin on his face, so I knew that I had scored a hit with the chief of staff,' says John. 'But it was probably not a career-advancing one.'

22. FUNNY YOU SHOULD SAY THAT...

Considering the dangers they faced and the conditions under which they had to work, it is remarkable that Sappers maintained their sense of humour. But sometimes it was only their ability to have a laugh and share a joke that got them through the most trying circumstances, not least the consequences of their serious shortcomings in basic soldiering skills.

'Most of us tend to think that the splinter teams were used by the battalions as de facto grunts—which we were,' says Allan 'Blue' Rantall of 1 Field Squadron. 'But I am not sure they got great value out of me. I remember once patrolling out somewhere dark and green when a five-minute halt was called. I promptly went to sleep, only to be awoken by a vaguely heard "Let's go!". Still only half-awake, I headed off in a different direction to the rest of the platoon.

'Another time, somewhere out in swampland, I went arse over tit after tripping on a tree root. My rifle ended up sticking out of the mud in a perfect vertical stance, with the muzzle six inches into the mud. You could have stuck a slouchie on it and it would have made a good marker. But I wasn't as bad as the forward scout who was carrying an F1 sub-machine gun and tripped like I had. I don't know how he did it, but he actually managed to shoot himself in the arse. Explain that one to the folks back home!'

Sometimes the trouble in the ranks was caused more by design than by accident. 'When we got back from one operation,' recalls Allan, 'the whole of 1 Troop was swung into action to build a massive Armco culvert across the full width of the newly built Kangaroo helicopter pad. We finished this just in time for Christmas, but just before the big day one of the planties let off a CS grenade that spread throughout the whole of the squadron. Nobody would own up to the deed so the officer commanding declared a dry camp until the culprit put his hand up. We hung out for two days, but as Christmas was coming he finally gave himself up and thus prevented a major catastrophe.

'Christmas Day started off with the 1 Troop's commanding officer, Mike McCullum, coming around with Brett Nolan and giving us "coffee royale" as we lay in our beds suffering hangovers from the night before. Then, later in the morning, it was off to the Luscombe Bowl to see a show featuring Lucky Starr, Lucky Grills and Lorrae Desmond. The highlight of the show was when one of Lorrae's boobs fell out while she was doing a dance—it was great! Years later I met up with her at a function and was indiscreet enough to remind her of it. She is still embarrassed about it all.'

Some of the fun and games Allan encountered had a serious, even deadly aspect. 'Sergeant Brett Nolan loved mucking around with explosives, and at one stage he was devoting a bit of time to see if he could make the M16 mine jump higher,' recalls Allan. 'He was pretty keen on this and had even gone to the effort of creating his very own subterranean workshop in the bunker next to the troop store.

'It was not unusual to see him opening up a mine with a hacksaw. However, this all came to a sudden halt one day when, for a reason that will remain a mystery, smoke started billowing out of the bunker, followed quick smart by a very nimble Brett Nolan. Just as well, for the next moment the whole bunker just exploded, sending dirt and debris all over the place.'

Not long after this, Brett's tour was up. He was replaced by Col Campbell, who soon made his own mark on the troops. Col didn't like all the explosive and paraphernalia that was spread throughout the tents—which was pretty typical of 1 Field Squadron—so he decided to have it all cleaned up. 'I was one of those put on the detail to go around with a wheelbarrow and collect all the slabs of C4 and det cord stored under the stretchers,' Allan says. 'By the time we had finished, we had collected four wheelbarrows full of the stuff.'

Then Col really set about making his name. 'A few of us were put to the job of raking up all the dead leaves that had piled up around the lines,' Allan continues. 'We had the job done just before lunch, so Col told us to set fire to the piles before we went to the mess—this we obediently did. Unfortunately, while we were at the mess, a bit of a breeze sprang up and the fires spread across the path to the Plant Troop lines. Before anybody could react, four of the tents—with all their gear—had burnt to the ground. Ah well, the planties were getting a bit uppity anyway!'

Corporal Peter Aylett had started as a plantie with 17 Construction but, as we related earlier, when he went back to Vietnam for a second tour he found that his NCO status had elevated him to the MATT, where his skills as a field engineer were soon put to the test.

'The first time I tried to blow the end off a 155 canister, I didn't know what I was doing,' says Peter. A canister is an artillery shell designed to explode and scatter shrapnel on impact; Sappers frequently had to disarm or destroy unexploded ordnance of this type. 'I thought I'd make it safe, and I went right across the road from the compound and dug a hole in a paddy field. I put some det cord under it to blow the bottom out of it and it shot out like a bloomin' rocket—I was glad no one saw that faux pas.'

Like many Sappers, who, if nothing else, are fundamentally practical and pragmatic soldiers, Peter had his share of run-ins with senior officers—or, at least, with those who assumed that their superior rank brought with it a wisdom that, in fact, only months in the field could bestow.

'We had a new warrant officer come into MATT 9,' recalls

Peter. 'We'd been without our original leader for a few months—he was a good bloke but a bit of a fusspot, and he started getting ratty so we sent him to Vung Tau for some R&R. So we got this new bloke, a tankie. We were going out on an operation and this new bloke said he'd organise it.

'That was fine, until I heard he was ordering choppers with 12 men from the South Vietnamese Army in each one. I said, "Listen you can't put 12 men in a chopper," and he said, "Why not? They're only little blokes." I said, "Have you seen how these guys travel? They take everything with them—they even carry their own water for the duration—they don't get re-subbed. By the time they're loaded up, they're as heavy as any digger with a machine gun." I told him it should be eight men at most. He said, "Listen, I'm the warrant officer and there will be 12 per chopper."

'I knew the chopper pilots wouldn't take them and we'd look bloody stupid with half the company in the air and the other half sitting on the ground waiting to take off. And once they'd done the sortie they'd be off somewhere else. They wouldn't come back for the rest.

'So I went round to the South Vietnamese company commander and told him, in my best Vietnamese, what was going on. I asked him to have a quiet word with the warrant officer and to diplomatically suggest that he rethink his plans. Anyway, the next thing I knew, he came round the corner and was right into him, yelling, "Warrant Officer, you cannot put 12 Vietnamese soldiers in one helicopter!" So the leader knew exactly what had happened. He pulled me into the hut and was reading me the riot act, pushing me and prodding me.

'I told him to calm down, but that just made him worse. I gave him three warnings, and he had his right index finger two inches from my face so I got it in my mouth and bit down as hard as I could. Well, then he really went off. He was on the radio to headquarters and broke all sorts of security protocols by mentioning my name and saying I wasn't suited for MATTs. So I got charged... and quite rightly—I had bitten a warrant officer's finger. Then the jokes started going around. "Did you hear that a warrant officer in MATTs got rabies? He got it from an engineer corporal."'

Peter was called before Ken Phillips, a major in the tank regiment. 'I had just bitten a tankie warrant officer's finger, so I knew I was in real trouble,' says Peter. But Major Phillips was quite intrigued when Peter explained how half the company would have been left sitting at base. He agreed that Peter was right but said he should have contacted him. Peter said he could not have done that without getting on the radio—breaking protocols—or without informing his boss. 'Alright,' Major Phillips said, 'you did a bloody good job—but why did you bite his finger? Why didn't you just haul off and hit him?'

'Sir, I wanted to use the minimum force necessary to restrain him,' Peter replied.

'You're a smart corporal,' Phillips answered, 'but you've got to see my point of view. I can't have corporals going around biting warrant officers—you could start a craze.' Then he asked Peter if a reprimand would be alright.

'Yes, sir—thank you, sir,' Peter replied. 'Just put them with all the others that I've got!'

Shortly afterwards, Peter was transferred to the battalion of Captain Len Opie. The reason given was 'clash of personalities'.

The first AATTV teams were elite groups of officers and sergeants, but as the numbers increased for the MATTs, the level of seniority was downscaled to teams of warrant officers and corporals. As Peter recalls, however, the message didn't get down the line to everyone.

'There was only one sergeant that I knew of on the MATTs, and we'd been over there for about nine months when someone queried why he was not a warrant officer, as the other senior men were,' he says. 'It turned out that he had been promoted but had never been told—and he had never been paid the extra money. He had worked for those nine months and then suddenly he was getting paid the difference between sergeant and warrant officer in back pay. Wouldn't he have had a good time!'

When the Sappers weren't battling their senior officers or the bureaucracy, there was the small matter of the Vietcong to deal with. And, as Alex Skowronski of 17 Construction attests, they were a clever and quick-witted foe.

'Denis Quick was operating out of Xuyen Moc, where there was a section of APCs doing an operation with one of the Vietnamese companies,' Alex remembers. 'They came across a small clearing, in the middle of which there was a bomb of some description. Denis decided he should blow it up. So he got his demolition kit out, cut a fairly lengthy fuse attached to the det cord, lit it, ran back to the APC and waited... but nothing happened.

'He reckoned the fuse had burnt out, but being a good soldier, he waited the necessary time and then went back to investigate. He was approaching the bomb when he suddenly stopped. The bomb was still there but the explosives were gone! The Vietcong must have been sitting in the scrub watching him, and as soon as he took off they'd gone down into the clearing, grabbed hold of the safety fuse, cut it and took off with a pound of C4, a detonator, a length of det cord and a reasonable length of safety fuse. Denis's eyes lit up like big neon signs when he realised he'd been so close to a very crafty enemy.'

Trevor Shelley also had an encounter that was a bit too close for comfort when he was part of a splinter team investigating a tunnel. 'I'm not sure of the operation or who my number two was, but I am pretty sure this happened after Long Tan but before Christmas 1966,' he says. 'It was just a small village in the red-clay rubber country and we found a tunnel entrance. The company we were travelling with had set up a defensive position around the area, and the two Sappers with them went to work.'

Trevor wasn't fazed at the thought of going down tunnels; by this stage, he had encountered a few, and this one didn't look out of the ordinary. 'I was crawling along on my hands and knees, as it was fairly big,' he says. 'I was checking the walls and the floor for abnormalities when the tone coming out of the tunnel wall changed from a dull thud to something akin to the sound of a kettle drum. It was a kind of eureka moment.

I informed my number two what I was doing and began slowly scraping off the clay and mud from a surface that turned out to be flattened 20-litre cooking oil drums.

'I removed the panel and shone the torch inside, taking care not to expose anything other than the hand with the torch in it. Inside, I could see a large square room that was almost filled with a large container also made from flattened square drums. Later we found that it was full of polished white rice. There was no movement and no sound, so with my confidence growing, I worked my head and shoulders inside. It was at that moment that I sensed—or saw in the corner of my eye—something akin to a missile heading toward me.

'In an instant I pulled back, and a fighting cock decked out with clip-on spurs went sailing past me, only inches from my nose. Along with my admiration for the ingenuity of this "watchdog", I learned another important lesson: if you can avoid firing a 9-mm pistol in a confined space, you should do so.'

Being a lieutenant in 17 Construction, Roger Cooke saw Sapper life from a slightly different perspective—but learning from your mistakes applied just as much to those in senior ranks as it did to lowly Sappers. 'It was discovered that there was a shop in Cam My that sold beer,' he relates, 'and immediately our Sappernuity came to the fore. Each afternoon at the end of our shift, the road crew called into Cam My to do some civil affairs work such as upgrading the roads and building playgrounds for the schools. I'd always wanted to have a go at driving the fitters' track (which is like an APC with a crane on it) so off I went. I was reversing, though, so I was relying on the bloke in the cupola to tell me when to stop. Unbeknownst to me, the bloke in the cupola was yelling to stop but was not plugged into the intercom.

'I reversed into a house and half-destroyed it. I tried to change gears to go forward, but I accidentally stayed in reverse so I went further into the house and destroyed the rest of it. I got some chippies rushed over from FSB Garth and built a replacement house that was bigger and better than the original. From then on, whenever I went into Cam My, the villagers would shout, "Mr Uc, Mr Uc!"—*Ucdaloi* is Vietnamese for Australian—"You drive into my house!"'

He also remembers one particularly peculiar encounter. 'One day, when the mine-clearing tank was clearing the roadside, a Citroën car pulled up and a couple of civilians got out. One of the men said that he had designed the tank and asked how effective it had been. No one thought to ask how it might be that a civilian had designed it—and that he was in Vietnam.'

A more predictable encounter occurred when, one night, sentries in one of the strong points observed some Vietcong fighters moving in near the rock-crusher in the quarry. 'It was decided to open fire with the M60 machine gun to frighten them away,' Roger says, 'and then to pick them off when they were in the open. Things did not turn out as planned. The Vietcong did run into the open but our shots sprayed everywhere because the gun was being fired through a chain-mesh screen. Inspection the next morning revealed that the generator was riddled with bullet holes and required considerable repairs.

The report back to squadron headquarters the next day read: "VC and NVA engaged in quarry, one ADF KVA WIA." Roughly translated, the latter part meant: "One Australian Defence Force power generator wounded in action."'

Another unexpected visitor arrived when an American Chinook landed at NDP Garth unannounced. 'It was a travelling salesroom for the giant department store Sears Roebuck,' explains Roger. 'They were just passing by and thought Garth was an American base. The idea was that the Yanks could look at samples on the Chinook and place orders then and there; the items they purchased would be delivered to their home addresses. Needless to say, they did not do any business at Garth.'

Any road convoy between Garth and Bien Bah had to have an APC escort, but this did not apply to the Salvation Army. 'They would turn up at the roadhead out of the blue in their Land Rover with cold orange cordial and biscuits,' Roger says. 'Every returned serviceman has nothing but praise for the Salvos.'

However well trained they were, there were always a few gaps in Sappers' knowledge. 'Each morning, the engineers manned the strong points while the infantry made their dawn clearing patrols,' says Roger. 'One morning, a new boy was manning the strong point at the gate to Garth when a Vietcong soldier approached with his hands in the air. He was saying, "*Choi hoi*," which means "I surrender". The Sapper didn't know what to do so he just ignored the Vietcong, whose arms probably got tired as he waited. Eventually, he was about to return to the bush when the clearing patrol came on the scene and rescued the situation.'

The first thing any experienced soldier does when he gets to a new place is to establish his 'bivvy'—a sheltered spot where he can roll out his swag, protected from the worst of the elements... and from his mates. Sappers have an advantage, of course, because they have the skills and equipment to build stuff. Lance Corporal David Roper of 3 Field Troop and, later, 1 Field Squadron recalls setting up camp near the Barrier Minefield, using his trusty bulldozer.

'I had finished putting a bund around the camp,' he says, 'when I had this brainwave—I would dig a small additional slot with the dozer. A few of us hooked our hoochies together and had a little Taj Mahal! It was great until the second night, when it rained and we had our own little swimming pool. After that first fiasco, I set up not far from the Centurion tank. One night it was asked to provide fire support and aimed directly over me. The muzzle blast blew my hooch away, leaving me suspended in mid-air. The tankie was locking his hatch as I was out for blood.'

While a huge amount of planning and hard work goes into setting up an FSB—most of it carried out by Sappers—double that amount was required to dismantle it and move both men and machines to another location. In 1971 Captain John Tick was one of the last to leave FSB Beth, just west of Xuyen Moc; his men were to create and move to FSB Ziggie, which was near the destroyed village of Tua Tich. On the final day at FSB Beth, the last men standing were the pioneer platoon and John's small CET of four Sappers.

FUNNY YOU SHOULD SAY THAT . . .

'After filling in defensive positions, disposing of accumulated artillery duds and other unserviceable munitions, and filling in the garbage pit,' John says, 'the rear security group would depart by helicopter for the new FSB. Everything went to plan, so by early afternoon it was time for the rear infantry and Sapper security detail to depart.'

The transport for the extraction was an American CH47 medium lift helicopter, which had been working all day moving defence, POL (petroleum, oils and lubricants) and other heavy stores to Ziggie. It was the end of the dry season and the day was hot. The large number of helicopter missions had basically destroyed the landing zone's oiled surface, which was very dusty as the team wearily climbed on board and waited for take-off.

'With increasing pitch and screaming whine, the CH47 built up its engine revolutions and prepared to take off over the slight west-facing slope of the landing zone,' says John. 'The pioneers and Sappers cast a last glance towards Beth, but because of the huge amount of dust surrounding the chopper we could see nothing. Suddenly there was a violent *thwack, thwack* sound and leaves and branches started pouring into the chopper up through the floor winch aperture. The two big African-American gunners on either side of the chopper had eyes like dinner plates—clearly, the take-off was not going according to plan.

'The noise in the chopper reached a crescendo of screaming engine noise and cyclical thwacking as the chopper blades cut their way through the rubber plantation. After what seemed like an eternity but must only have been about ten seconds, the chopper crash-landed about 50 metres into the rubber plantation, surrounded by smashed trees and leaf confetti and with the engine whine screaming but decreasing.'

On board, the stunned Sappers and pioneers wondered what the hell had just happened. The question of what to do next was soon resolved by the US crew gunners, who were shouting, 'Get out and run like hell! Run! Run! Run!'

'Funnily enough, the pioneers and Sappers took off in a fairly orderly fashion, but we evacuated the chopper in seconds,' John remembers. 'We stopped about 50 metres to the rear of the chopper, which by now had its engines almost wound down. The interlude was disrupted by the US chopper crew, who were racing further to the rear and shouting, "Run! Run! She'll blow!"

'In seconds, the world sprint record was broken by 28 Aussies and five US servicemen running like hell. Once about 100 metres from the chopper, everyone stopped and looked back at the wreck, fully expecting it to explode in a fireball. It looked like a mortally wounded bird lying on its chest with wings spread. The rotor blades were shredded—only the rotor stubs were slowly rotating.'

Eventually, the US crew walked cautiously back to the chopper and radioed their command, calling for a recovery mission to be sent. The Aussies sat around marvelling at their survival and having a brew, and an hour or so later a huge Sikorsky Skycrane appeared. Heavy lift straps were tethered to the CH47, which was then lifted clear of the plantation and flown home to base.

'For months after this event,' John concludes, 'the Sappers of 2 Field Troop were, to a man, very wary of a certain CH47 that exhibited an excessive vibration when flying under load.'

From carrying huge bridge beams and dropping them into place, to rescuing smaller helicopters, it seemed there was nothing the Skycrane couldn't do. But as Second Lieutenant Jim Straker of 17 Construction discovered, the crew of the giant helicopters sometimes had different priorities from those of the people on the ground. A new kitchen was to be delivered to the Task Force headquarters at Nui Dat. It comprised three prefabricated units—each ten metres by three metres—which would be lifted in by Skycrane one at a time.

'The Sikorsky has a main hook that can be opened electrically, and four ancillary hooks that have to be opened manually,' says Jim. 'Unit 1 was to be delivered using all five hooks so that the load was carried in a controlled manner; the manual hooks were to be opened by ground staff, who would drive to the load, climb a ladder up to the hook and then move clear.'

The Sikorsky duly appeared in the sky and headed to Luscombe strip to offload the units; from there they would be taken to site on a low-loader. Unfortunately, Luscombe was being extended at the time, and there was a lot of loose material on the ground.

'There was an almighty cloud of red dust around the unit as it came close to the ground under the 100 miles per hour downdraft from the Skycrane,' says Jim. 'As the pilot moved up and down the strip, trying to find a clear area, a team followed it in a Land Rover—it was a bit like in *Keystone Cops*. Eventually, the verdict was "no go" and so the helicopter and its load were directed to return to Vung Tau.'

As the pilot flew over Nui Dat, he spied the helipad, which was all bitumen-coated and so was unlikely to cause a dust nuisance. However, it was also in constant use by tactical choppers. 'Over the hill he went,' Jim continues, 'and landed the unit in the middle of the pad. The loadmaster climbed down and undid the external hooks. When the Land Rover arrived, the pilot beat a hasty retreat, leaving the engineers from 17 Construction to bring cranes and a low-loader onto the middle of the helipad to remove the unit so that combat choppers could land and take off and the war could continue.

'Unit 2 was dispatched the next day. The Americans had learnt from having to deal with four external hooks, so this one was rigged using only the central hook. That was great in theory but caused some distress to the helicopter and the loadmaster when the unit moved during flight and broke into the loadmaster's bubble.'

Finally, the delivery of the third unit saw them get it right: they kept the unit under control by using a drogue parachute. 'The kitchen was put in place and the Task Force headquarters settled down to better food,' says Jim.

23. THE FINAL DAYS

Australia's involvement in the Vietnam War started to wind down in November 1970, with the Australian government keen to reduce its commitment and to put the responsibility for South Vietnam's defence back on the Vietnamese themselves. Eight RAR was not replaced at the end of its tour of duty, meaning the Task Force was reduced to just two infantry battalions, with armour, artillery and air support. The Australian forces were spread more thinly across Phuoc Tuy province, although sustained actions between September 1969 and April 1970 had seen the communist activity in the area reduced; it was believed that many Vietcong and North Vietnamese forces had withdrawn, if only to recuperate. Despite occasional local actions by the remnants of the Vietcong, Highway 15, the main route running through Phuoc Tuy between Saigon and Vung Tau, was open to unescorted traffic for the first time in years.

But the North Vietnamese hadn't given up—far from it. Australian combat troop numbers were reduced even more during 1971; the Battle of Long Khanh, which took place in early June 1971, was one of the last major joint US-Australian operations. It resulted in three Australians killed and six wounded. On 18 August 1971, Australia and New Zealand decided to withdraw all their troops from Vietnam, with Australia's prime minister, William McMahon, announcing that 1 ATF would cease operations in October, which would lead to a phased withdrawal. The Battle of Nui Le on 21 September would be the last major battle fought by Australian forces in the war; it resulted in five Australians killed and 30 wounded.

Sappers were, of course, in the thick of the action right to the bitter end. The Australian forces firstly withdrew to Nui Dat, which had come under the defensive protection of 4 RAR, and gradually the rest of the forces were moved south to Vung Tau for return to Australia. But Sappers were still providing mini-teams and splinter teams for the patrols that continued to be vital to the security of the Nui Dat base. Australian forces would never be more vulnerable than during a mass withdrawal.

Four Sappers, under Corporal Rod Richards, stayed with 4 RAR as the remnants of the Task Force headed for home. In an echo of the Sappers' first involvement as a unit in Vietnam—Sandy's 3 Field Troop—this last vestige of the engineer presence was named 3 Independent Field Troop. When the main body of 4 RAR moved to Vung Tau on 16 October, having handed over control of the base at Nui Dat to South Vietnamese forces, the Sappers were assigned for the very last time in mini-teams to A Squadron of the 3rd Cavalry Regiment. The heavy convoy of 1 Field Squadron would be the rear party of the withdrawing forces.

Finally, 4 RAR, the last Australian infantry battalion in South Vietnam, sailed for Australia on board HMAS *Sydney* on 9 December 1971. Australian advisors with the AATTV remained in the country, however, and continued to train Vietnamese troops until the newly elected Australian Labor government of Gough Whitlam decreed that the last of them would be withdrawn by 18 December 1972.

In total, approximately 60,000 Australians—ground troops, air force and naval personnel—served in Vietnam between 1962 and 1972. Five hundred and twenty-one died as a result of the war and over 3000 were wounded. Some 15,381 conscripted national servicemen served from 1965 to 1972, of whom 202 were killed and 1279 were wounded. Between 1962 and March 1972, the estimated cost of Australia's involvement in the war in Vietnam was $218.4 million.

Just under 1600 Sappers served with 1 Field Squadron in Vietnam—many engineers served with other units but it was 1 FS that suffered the vast majority of casualties. In all, 36 Sappers were killed in action. A large part of the death toll can be attributed to the dangerous work they did defusing, laying and clearing mines. But the Sappers had double duties, backing up with infantry and armour, and often leading the way through minefields, booby-trapped bush and tunnels. They *did* make and break, and they *were* everywhere, pioneering the methods that are standard practice for Sappers in war zones today.

Vietnam changed everything for army engineers. It truly was a Sappers' war.

AFTERWORD
SAPPERS SALUTED

Six Sappers had their valour recognised during the Vietnam War by the award of the prestigious Military Medal. The Military Medal is not given lightly. It's about valour and bravery under fire. These are the official citations that accompanied the Military Medals awarded to the six Sappers in Vietnam.

Lionell Rendalls MM

On 17 October 1966, during Operation Queanbyan, Corporal Rendalls was in charge of an engineer combat team in support of an infantry company. The company, whilst moving into a village area, suffered seven casualties from booby traps. Corporal Rendalls moved to the head of the company, and covered by only one scout to his rear, proceeded to clear forward. During the next two hours, working alone, he located and neutralised seventeen booby traps including a type previously unknown. Corporal Rendalls, during this clearance operation, displayed considerable skill and unquestioned bravery. His efforts permitted the Infantry to continue their operation without further casualties, and the calm manner in which he performed this arduous task reflects great credit upon himself and his Squadron.

Neil Innes MM

On 2 May 1967 during Operation Leeton, Sapper Innes was in the 1 Troop forward operational base at the Horseshoe. At approximately 1150 hours Sapper Innes heard an explosion in the vicinity of a minefield approximately 70 metres to the south of his own location. He immediately ran to the edge of the minefield where he saw two American Gunners lying wounded on the ground. One soldier was unconscious just outside the minefield, the other was lying in the minefield and was writhing on the ground approximately six inches from another mine. Sapper Innes entered the minefield, held the wounded soldier firmly to prevent his movement and quietly reassured him to lie still. He then marked a clear lane into the minefield to allow medical aid to come forward. Sapper Innes, by his immediate and courageous action, prevented the detonation of a second mine and his swift marking of a safe lane made possible the quick evacuation of the wounded. His complete disregard for his own safety displayed a high standard of bravery that reflects great credit upon himself and his Squadron.

AFTERWORD

Ray Ryan MM

On 10 March 1969 Sapper Ryan was the team leader of an engineer splinter-team which was allotted to D Company, 5th Battalion, The Royal Australian Regiment. At about 0200 hours, elements of the company entered a minefield and suffered a number of casualties from mines and small arms fire. Sapper Ryan moved forward and commenced to clear paths to the wounded. He worked single handed in the darkness for four hours, prodding through the minefield with a bayonet and a torch to reach the casualties. After he had reached the last of the wounded and was withdrawing from the minefield he detonated a mine and was badly injured. He was subsequently evacuated to Australia as a result of the wounds he received. Sapper Ryan displayed great coolness, bravery and distinguished conduct both before and after he was wounded. His actions reflect outstanding credit on himself and his corps.

Ronald Snow MM

From June 1969 to September 1969 Sergeant Snow commanded the 1 Field Squadron Land Clearing Team during the clearing of the jungle from foothills of the Long Hai Mountains. During this operation, Sergeant Snow and his men, dressed in hot, heavy flak jackets and steel helmets operated armoured TD 15 Bulldozers throughout the monsoonal period, clearing the jungle in difficult terrain. They were frequently attacked by enemy firing at their bulldozers with anti-tank rockets from almost point blank range. With their bulldozers they also detonated more than sixty landmines that were laid by the enemy in an attempt to deny access into the area. While operating the armoured bulldozers machinery within the jungle, Sergeant Snow and his men had restricted hearing and vision which made them vulnerable to enemy attack. Sometimes they were not immediately aware they were under attack and a number of plant operators were wounded.

This sustained heroism along with the outstanding results achieved—with more than 4000 acres of jungle being cleared—was largely due to Sergeant Snow's personal bravery, initiative and drive. He inspired his men who were undaunted by the casualties they suffered, the dangers to which they were exposed or the conditions under which they operated. His example was in the finest traditions of his corps and reflects great credit on himself, his squadron and the Royal Australian Engineers.

Phillip Baxter MM

On 21 July 1969, Corporal Baxter was the Commander of an engineer mine clearing team supporting 3 Platoon, A Company, 6 RAR. While the platoon was deploying into a harbour position, a mine was detonated killing the platoon commander and wounding sixteen others, including Corporal Baxter. Though in pain and suffering slight shock from his wounds, Corporal Baxter calmly assessed the situation. He called on the radio for deployment of a combat engineer team to assist with ground clearance for further mines. He immediately began to clear and mark safe lanes to wounded personnel,

AFTERWORD

coolly reassuring the wounded and assisting with first aid where necessary, despite his own quite serious wounds. He continued the clearance with the assistance of another field engineer until all other wounded members had been treated, and then commenced clearance of a helicopter landing pad to allow evacuation to proceed. When the combat team arrived to take over further clearance, his own wounds were tended. In a situation where avoidance of alarm and minimum movement was essential to prevent additional casualties, Corporal Baxter's actions were exemplary, an inspiration to those about him and in the highest traditions of his Corps and the Australian Army.

Gary Miller MM

On 6 March 1970, Sapper Miller was the leader of a two man engineer team supporting a rifle company of 8 RAR and when an enemy mine was initiated by a member of the force. Despite the danger to himself, Sapper Miller immediately set about the hazardous task of clearing safe lanes by hand to allow access to the casualties from the incident. Undaunted by the knowledge that the enemy normally lays mines in lots of two or more in an area, Sapper Miller meticulously cleared the necessary avenues of approach to the wounded and then proceeded to clear an area from which the casualties could be winched out of the area by helicopter. His complete lack of fear was an inspiration to those who we represent. His appreciation of the situation led to additional engineers being flown into the area.

A SAPPERS' WAR

On 15th March 1970, Sapper Miller was again supporting an infantry force when a mine was initiated, resulting in four persons being wounded. Heedless of his own safety, Sapper Miller again undertook hazardous hand clearance of the area. Having cleared safe lanes into the casualties and cleared the remainder of the area for movement, Sapper Miller was not satisfied that he had done all that he could. Bearing in mind the enemy mine practice, he relentlessly searched the surrounding area in an effort to find the second or more of the mines. His pursuit of his objective was rewarded when he detected another anti-personnel mine sixty metres from the initial explosion. This mine was found in an area through which his or a similar force would have undoubtedly passed. It is doubtless that this action prevented further casualties. In both instances, Sapper Miller's fearless and relentless application to his task was an inspiration to those around him and was a magnificent example to all.

Other Australian decorations for Sappers in Vietnam include:

Military Cross
Sandy MacGregor and Brian Florence

OBE
Colin Browne

AFTERWORD

Mentioned in Despatches

Terrence Raymond Binney, Robert James Earl, Robert Walter Fisher, Ronald Percy Janvrin, Raymond James Johnson, Phillip James Jones, Graeme Ernest Leach (aka Graeme Arthur Patrick Eggins), John Fullarton Meldrum, Walter Elwyn Rogers, Rex Rowe, Murray Clarke Walker and Henry P. Zurakowski.

PRAISE INDEED

When all the inter-service and inter-unit rivalry is removed and the tub-thumping and chest-beating is toned down, servicemen respect each other because only they can really know how hard their jobs are—to be separated from friends and family and, in the case of war, to face death and injury every day.

Sappers often aren't shy about blowing their own trumpets—at least in the company of other soldiers, sailors and airmen. But what do other soldiers think about Sappers? *Holdfast* online newsletter collated some views expressed by the Vietnam War's senior Australian officers, and this is what they said.

PRAISE INDEED

Brigadier Colin 'Genghis' Kahn DSO, former Commanding Officer of 5 RAR on their second tour (February 1969 to February 1970)

Whenever 5 RAR operated, be it in Long Khanh, Bin Tuy, Bien Hoa etc, we hit bunker systems ranging in numbers from 6 to 600 bunkers. The call went out continuously—'We need the Sappers', and the wait for the Splinter Teams was not long, for they were always up there, forward with the infantry.

5 RAR located some 1500 bunkers in the period and the need for Sappers was great. The Splinter Teams attached to, moving and fighting with rifle companies, searched these bunkers for mines and booby traps, and when we left a bunker complex, they either destroyed them or rendered them uninhabitable with CS crystals. There are numerous riflemen from 5 RAR who came home safely because of the support, the courage and the professionalism of our Splinter Teams.

I believe, and I have always believed that the Sappers who moved and fought with Infantry and Armour deserved some distinguishing recognition such as a 'Combat Engineer Badge'. We Infantrymen have our Combat Badge—Splinter team members deserve something similar.

Major General Mike O'Brien, former Platoon Commander and Intelligence Officer with 7 RAR on their second tour— February 1970 to February 1971

No infantry platoon, company or battalion commander could have done without his engineers. As always, war was a team effort, and each part of the specialised team was essential if the team were to succeed. The Sappers were always essential and particularly so in the post-minefield part of our Vietnam War. We rarely gave them the credit due: we often did not allow for their redeployment to and from our companies, the need to know who was who and the need to settle into a team. We demanded high performance straight away—but we always got it.

They developed a particular expertise suited to our specific operational methods: their appreciation of mines and booby traps was often little short of uncanny. They had distinctive personalities, perhaps less regard for the niceties of neatness than some, but when it came to the crunch, they performed and performed superbly. They saved the lives of many an infantryman and we should be eternally grateful. If there is a sad note, it is that few infantrymen realise that it was the Sappers in Splinter, Mini and Combat Engineer Teams from the Engineer Field Troops that suffered our highest casualty proportion in the Vietnam War: a grim testament to their effectiveness. Look at their list of killed and wounded and remember it next time we say 'Lest we forget'.

3 RAR on their 1967-1968 tour had nothing but praise for the Tunnel Rats who worked with them

In most Battalion operations, the Engineer Combat Teams of 1 Field Squadron were either an integral part of an individual rifle Company or were held in readiness at Battalion headquarters to be inserted in support of Companies when called for... The task for which they will be most remembered by the Battalion however, was their task of clearing cunningly concealed and potentially lethal mines and booby traps.

3 RAR's Operation Pinnaroo during March-April, 1968 involved penetrating into the Long Hai mountains, an area peppered so heavily with mines, minefields and booby traps that the Tunnel Rats from 3 Troop 1 Field Squadron led the advance that took many days. It was a costly Operation, with casualties from mines and booby traps far exceeding casualties from small arms fire. Task Force Commander Brigadier R L Hughes, DSO sent the following message after the Operation: 'I want to commend everyone who participated in Operation Pinnaroo for their steadiness and professional ability during an Operation which I believe has been the most difficult and dangerous yet undertaken by 1ATF. Particularly I want to commend the Engineers and Pioneers for their work in the minefields. The completion of Pinnaroo is a job well done.'

On 24 April, 1968, 3 RAR celebrated Kapyong Day and 3 Troop 1 Field Squadron was not only invited by the CO Lt Col J J Shelton, MC to parade with them, but was drawn up in front of the Battalion. It was a significant and unique honour for the Sappers.

Colonel Peter Scott, former Commanding Officer of 3 RAR, Vietnam, 1971

The existence of the Vietnam Tunnel Rats Association came to my attention through the article 'Battle of Long Khanh' [published in *Holdfast*]. It was a great article and it included quite rightly the contribution of the 'Splinter' Teams of 2 Troop,

1 Field Squadron attached to 3 RAR for that operation. There were many photographs taken by the then Captain John Tick, OC 2 Tp, which I saw for the first time.

The article brought home to me the embarrassing fact that I had never acknowledged the great job the sappers of 2 Tp did in supporting the battalion during the whole of our tour of Vietnam in 1971. It's very late but I do it now ... In Vietnam we visited the 1 Fld Sqn 'Mines Room' during our 'In Theatre' training and this alerted us all to the dangers of mines likely to be encountered in Phouc Tuy Province. However the real advantage to 3 RAR was to have the Splinter Teams attached to each company for every operation. Without examining the actual time spent on operations I would estimate the companies were on operations for about 90% of the time in Vietnam. This meant that the 'Splinter' Teams and members of 3 RAR had plenty of time to know each other and gain each other's respect and confidence.

The diggers of 3 RAR relied heavily on the advice and confident way in which the 'Tunnel Rats' undertook their tasks which, were many and varied. Destroying UXBs and bunkers were the most prominent and frequent of these. The bunkers encountered during the Battle of Long Khanh are an example of the enormous task that the members of 2 Troop undertook willingly and with great professionalism under very difficult conditions. On behalf of the members of 3 RAR who served in Vietnam in 1971 I belatedly congratulate all the Tunnel Rats who provided such outstanding support during our tour.

THE WAR THAT NEVER ENDS

In the old days, they called it 'shell shock'—the often debilitating and damaging effects of having experienced things in war with which many of us, if not most, are simply not equipped to deal. These days they call it 'post-traumatic stress disorder' or 'PTSD', although there are many still in denial about its effects. During the research for *Tunnel Rats*, we were interviewing one of the originals who'd been with 3 Field Troop. Asked about PTSD, he scoffed, 'It's all bullshit, mate. Just some bludger looking for a pension.' As we absorbed his cynical assessment of the issue, his mate—both in the tunnels and in civvies—piped up, 'Of course, he has been married five times.'

All joking aside, according to a number of official reports, it is now clear that service in Vietnam had a long-term impact on many veterans' health—particularly their mental health—and that of their families. One study found that Vietnam veterans had a significantly higher suicide rate when compared with the average Australian male population. In another, the Vietnam Veterans Health Study, 30 per cent to 45 per cent of veterans reported suffering from particular mental disorders, while one in three vets in the Veterans and War Widows Study reported that their doctor or health professional had raised alcohol abuse as a possible health concern.

According to the Vietnam Veterans Association of Australia, the symptoms of PTSD can include tension and agitation; sleep disturbance, including dreams and nightmares; 'flashbacks' or intrusive memories and feelings; emotional detachment or 'coldness'; social withdrawal; self-preoccupation and/or egocentric behaviour; irritability; avoidance of reminders associated with trauma; mood swings; depression; anxiety or panic attacks; fearfulness; continual alertness for future emotional or physical threats; physiological reactions, such as headaches, stomach upsets or rashes; poor concentration or loss of confidence; and alcohol and other drug abuse.

Earlier in this book, George Hulse described an incident in February 1969 in which Sappers under his command had to repel an attack by enemy soldiers. One of the four Vietcong was killed, while the other three were seriously wounded. An Australian Sapper, identified only as 'M', was traumatised by the incident, although it would be 40 years before he finally sought help. He is just one of an army of 'invisible' casualties, and his story is illustrative of a much broader problem.

'The effects of that incident—now commonly understood

as PTSD—began to affect Sapper M,' says George. 'Almost forty years later, his PTSD had become so acute that he needed assistance from the Department of Veterans' Affairs (DVA) to help him cope with his invasive memories of that event. But the DVA needed proof that his condition was linked to his war service, so I was asked to write a letter detailing what had happened.'

This is what George wrote:

The aim of this testimonial is to assist Mr M's memory of events concerning his contribution to the war in South Vietnam (SVN) in the period July 1968 to July 1969. I was M's Troop Commander from January 1969 to Early April 1969. I was at that time 213644 Lieutenant George Leslie Hulse. My responsibility was to command the Plant Troop of the 1st Field Squadron Group, Royal Australian Engineers (1 Fd Sqn RAE).

The role of 1 Fd Sqn RAE during the war in SVN was to provide a range of combat engineering support to combat units of the Australian Task Force based at Nui Dat in Phouc Tuy Province of the then SVN. We were called 'Combat Engineers'. The role of Plant Troop of 1 Fd Sqn RAE was to construct mobility resources in support of combat operations and to maintain as much of the civilian infrastructure as our resources were able to provide. Our combat objectives included the need to provide roads, tracks, helicopter landing zones, land clearing operations, fortification of strong points such as isolated forts, the construction of combat facilities such as excavation with machines of deep holes for the emplacement of Headquarters and Regimental Aid Posts,

A SAPPERS' WAR

and the bunding of Artillery gun positions and Armoured Corps tanks.

This brought all sappers of the Plant Troop into close contact with the enemy. There were many occasions when the sappers of the Plant Troop were engaged in an infantry role where the requirement was to close with and kill or capture the enemy. During the night of 10 February 1969, Sapper M was engaged in a night contact with a squad of local force Viet Cong. When the contact was unavoidable, I gained authority from my superior commander (an infantry company commander) to open fire on the enemy soldiers and ordered my sappers, including Sapper M to open fire. Of the four enemy soldiers, we hit all four, killing one, and seriously wounding the other three. Although I congratulated my sappers for their contribution to that fire fight, I could see that Sapper M was not too impressed with his handiwork. He did not complain, but remained very quiet about that incident.

In addition, Sapper M was mainly engaged as a grader operator. This required him to operate behind the road construction/ maintenance crews in finishing off rough work into a smooth finish so that rollers could compact the now smooth road surface. Graders move more quickly than the other construction equipment, and there were times when the grader operators found themselves out of the range of the accompanying infantry and armoured protection. They were aware of their exposure, but worked on regardless of it. There were times when they came under fire, but, during my tenure as Commander of the Plant Troop, none of the grader operators were hit by enemy small arms fire. Nonetheless, it was a risk to the grader operator because he had no choice but to focus intently on his blade, his direction and his speed. He could not observe for enemy around him. Sapper M operated a grader for many months under these conditions.

Despite the pressure of operating in the open like a 'sitting duck', Sapper M never once complained or requested an easier job. When I needed volunteers for a mission, he would step forward without any coercion or coaxing. He never once feigned illness to get out of going to known trouble spots where contact with the enemy was a constant feature of that particular area e.g. the area contiguous with the Long Hai Hills and the 'Horseshoe Feature'.

Although Sapper M was a well accepted member of the Plant Troop and joined in during Plant Troop and 1Fd Sqn RAE social activities, I did notice, after the 10 Feb 69 contact, that he became a little more introspected than he had previously been. Not a total recluse, as his mates considered him to be a 'good bloke' (a definite compliment coming from plant operators), but not the same Sapper M as prior to that contact. I spoke to him in very general terms about that, but he assured me that it was all OK. I let it go at that.

I regret now that none of us had a better understanding of what was to become known as PTSD in the years ahead. I believe that had we been better educated in the phenomena of PTSD when I saw the change in Sapper M's personal behaviour, we may have been able to contribute more efficiently to his rehabilitation. Consulting my diary, 37 years after the event, and finding information to offer his current day rehabilitators, I realise is cold comfort for Sapper M's condition today.

A SAPPERS' WAR

I thank Sapper M for his selfless contribution to combat operations in a war at a time when Australia needed him. I hope that now that Sapper M needs Australia to help him, we will respond in kind.

Yours in soldiering
George Hulse
LTCOL (Retd)

'The letter materially helped, and M's condition was accepted by the DVA as being war-related, and he is now being assisted by them,' says George. 'His situation demonstrates the numerous incidents in respect of diggers involved in the war of South Vietnam, where the effects of the war were not recognised until many years later, when their conditions became unbearable for them and their families.'

It's a sobering thought. When all the bullshit and bravado dies down, when the fallen have been rightly remembered and the heroes properly honoured, there are those who carry scars from a war fought 50 years ago that will probably never heal. They need to know that it's never too late to ask for help, and that their comrades are still there for them—as they ever were.

For more information, please visit www.vvaa.org.au/ptsd.htm, or go to www.vietvet.org/vvcsaust.htm for a list of contact numbers for the Vietnam Veterans Counselling Service.

SAPPERS STILL CLEARING MINES

No one knows the devastating effect of landmines and unexploded bombs better than the Sappers. After all, they had to discover and defuse them, leading to 1 Field Squadron having the highest casualty rate of any Australian unit in the Vietnam War. Meanwhile, across the border in Laos, between 1964 and 1973 more than two million tonnes of bombs—including 260 million cluster bombs—were dropped during the 'Secret War' run by the CIA. That's more bombs than were dropped on the whole of Europe in the Second World War, making Laos the most bombed country on Earth. Around 30 per cent of these bombs didn't go off; today, instead of professional soldiers having to deal with them, children and other civilians are suffering death and terrible injuries.

The dry season 'crop' in Laos involves the families of

subsistence farmers looking for metal to sell to any one of the 15 foundries in Laos, where there is no iron ore to be mined. The foundries provide the metal detectors (bought in Vietnam for about $12) and the kids dig up the metal. All too often it is unexploded ordnance, and as a result, around 26,000 Laotians have been killed since 1975. About 300 more die every year.

That's why former Australian Sappers and other Vietnam veterans formed the MiVAC (Mine Victims and Clearance) Trust, which is a totally voluntary organisation set up to combat the problem of landmines in Vietnam, Laos and Cambodia. Membership of the trust has broadened to include ex-service personnel from other conflicts, humanitarian aid workers, members of peacekeeping forces and concerned civilians. It has been operating in South-East Asia for ten years, and more than 95 per cent of the money the trust raises is spent on projects there.

MiVAC is now focusing its efforts on Laos, where it clears unexploded ordnance (UXO) and minefields and runs other humanitarian projects. MiVAC has been working in the Xieng Khouang province of Laos. Following the clearance of the Khangpunghor 1 and Khangpunghor 2 minefields in October 2010, paid for by MiVAC, members carried out an extensive needs survey in the villages adjacent to these minefields. MiVAC is determined to address the extreme poverty and development needs of people in Ban Xai and its neighbouring village of Hinmu Pueng.

The 700 inhabitants of Ban Xai and Hinmu Pueng have

no clean water sources, no sanitation, no power and a school with a dirt floor, and face gross food-insecurity for most of the year. Using animal fertiliser for market gardens is a virtual unknown in Ban Xai. Most inhabitants are below the Laos government poverty measure of 50 cents per day, and many families live on as little as $30 per year, with a food intake of less than 2100 calories per day. The problems are compounded by the inflow of Hmong people relocated by the Laotian government programs. The Hmong race has been a persecuted minority, having been the main fighters used by the CIA during the Secret War and then abandoned to their fate.

MiVAC is working with Laotian authorities and with the people themselves to help eradicate the terrible threat of unexploded devices, while at the same time raising living and education standards among the nation's poorest people. One example of their work and its knock-on effect came in 2009, when a MiVAC team stopped in Ko Hai village for morning tea—en route to another project—and ended up clearing school playgrounds where four children had previously died; the team removed 84 UXOs. With the playground safe, more students enrolled, which led to another MiVAC member returning to the school to build accommodation for 80 day students, a water tower and new toilets. Now the school is trying to be more self-sufficient, planting corn and fruit trees in areas previously too dangerous even to walk through.

MiVAC is also helping to train and equip locals to become de-miners themselves. It seems only right that the people who knew these weapons best, whether from laying them or clearing them, are now instrumental in protecting innocent civilians from their deadly legacy.

For more information, please visit www.mivactrust.org.

Sandy MacGregor
Patron

SAPPER SOURCES AND OTHER READING

Quite simply, this book could not have been written without the assistance of the many veterans of the Vietnam War who shared their stories with us and, notably, two newsletters that have gathered, collated, edited and published Sappers' tales. In so doing, they have created a great repository, not just of historical detail but of the spirit of the Sappers they represent.

Holdfast is an online newsletter published, edited and—to a large extent—written by Jim Marett, himself a Tunnel Rat and now, as he describes himself, the 'Grand Poobah' of the official Vietnam Tunnel Rats Association. A lot of the information and some of the stories in this book have come directly from or indirectly via *Holdfast*, as have some of the pictures. Jim has kindly given us a direct link to the Tunnel Rats and their yarns. Soldiers may be great at telling a story but sometimes aren't the best at writing them down, and it has to be acknowledged that, at the very least, part of any of the stories that are attributed to Jim or are told as quotes from him have probably appeared in *Holdfast* and, as such, remain his copyright.

We are lucky to have Jim and his publication around as a source for some of the material. You can—and should—read the entire run of *Holdfast* by going to www.tunnelrats.com.au and clicking on 'Newsletters'.

But not all Sappers were Tunnel Rats, and another great source of engineer yarns is the *Phuoc Tuy News*, the newsletter of the R.A.E. Vietnam Association. We are grateful to its editor, Vincent P. Neale, for the stories (many written by himself) that he has allowed us to use, at least in part. You can read many more Sapper stories from Vince and others at www.raevietnamassociation.org/news.htm.

Meanwhile, other notable contributors who gave their time and their memories to this book include Peter Aylett, Phil Baxter, Ventry Bowden, Bruce Cameron, 'Stiffy' Carroll, Joe Cazey, Roger Cooke, Bob Coker, Bob Hamblyn, Les Hillberg, George Hulse, Neil Innes, Peter 'Pedro' Kelly, Peter Krause, Barry Lennon, Peter Lofthouse, Chris MacGregor, Clive Pearsall, John Peel, Alan Preston, Allan Rantall, David Roper, John 'Speedie' Sahariv, Doug Sanderson, John Schumann, Trevor Shelley, Alex Skowronski, Jim Straker, Dave Sturmer, Bob Summerville, John Tick, Ralph Todd and Rob Woolley.

Major Brian Florence is completing a meticulously detailed account of the Sappers' actions in Vietnam for publication in the near future. It covers the history, operation by operation,

SAPPER SOURCES AND OTHER READING

of the Australian Army Engineers in South Vietnam between 1962 and 1972.

The story of Bridge 6 is recorded in a 17 Construction Squadron scrapbook now held in the RAE Museum, Steele Barracks, Moorebank NSW. A documentary, by Summerlands Productions, is held in the National Vietnam Veterans Museum.

John 'Jethro' Thompson has written a book about his terrible injuries suffered during the laying of the Barrier Minefield and his inspiring life thereafter. Called *A Vietnam Vet's Remarkable Life*, you can find a video about it on YouTube here: http://youtu.be/ctPd-f0b3ic.

GLOSSARY

AATTV	Australian Army Training Team Vietnam
Acco	An Australian-built truck
ALSG	Australian Logistical Support Group
AO	Area of operations
APC	Armoured personnel carrier
ARV	Armoured recovery vehicle
ARVN	Army of the Republic of Vietnam (South Vietnamese Army)
ATF	Australian Task Force
Bund	Defensive earthwork wall
C4	American plastic explosive
Cav	Cavalry (but in APCs rather than on horses)
CET	Combat engineer team
Charlie	Vietcong (from the call sign for VC—Victor Charlie)

GLOSSARY

CO	Commanding officer (normally of li[eutenant] colonel rank, say of an Infantry Battalio[n]
CS	Tear gas
Digger	Australian soldier
DVA	Department of Veterans' Affairs
FSB	Fire support base
Grunt	Infantryman
IED	Improvised explosive device (booby trap bomb)
Lance-Jack	Lance-corporal
LCT	Land-clearing team
Louie	Lieutenant
LSM	Landing ship medium
LZ	Landing zone
MacV	Military Assistance Command Vietnam—the US's unified command structure for all of its military forces in Vietnam, often confused with ...
MacV-SOG	Studies and Observation Group which was, in fact, a covert American 'special operations' unit
MATT	Mobile Advisor Training Team
MC	Military Cross
MCT	Mine-clearing team
MM	Military medal
Nasho	National serviceman
NCO	Non-commissioned officer, such as corporal, sergeant, etc.
NDP	Night defensive position
NVA	North Vietnamese Army

A SAPPERS' WAR

OC	Officer commanding (of a sub unit, say an Engineer Troop [a Captain] or an Infantry Rifle Company [a Major])
OHP	Overhead protection (on bunkers)
Plantie	Plant operator (e.g. bulldozer driver)
R&R	Rest and recreation (or recuperation)
RAASC	Royal Australian Army Service Corp
RAE	Royal Australian Engineers
RAEME	Royal Australian Electrical and Mechanical Engineers
RMO	Regimental medical officer
RPG	Rocket-propelled grenade
SAS	Special Air Service
Sitrep	Situation report
SLR	Self-loading rifle (standard issue Australian weapon)
Tankie	Tank crewman
UXB	Unexploded bomb
UXD	Unexploded device
VC	Victoria Cross
VC (2)	Vietcong
WO	Warrant officer